Diane Ackerman is the author of the bestselling *A Natural History of the Senses,* among many other books of non-fiction and poetry. Her essays on nature and human nature have appeared in *National Geographic, The New Yorker,* the *New York Times, Smithsonian* and elsewhere.

www.dianeackerman.com

F
b
r
L
b

Also by Diane Ackerman:

An Alchemy of Mind
Cultivating Delight
Origami Bridges
Deep Play
I Praise My Destroyer
A Slender Thread
The Rarest of the Rare
A Natural History of the Senses
A Natural History of Love
Jaguar of Sweet Laughter
Reverse Thunder
On Extended Wings
Lady Faustus
Twilight of the Tenderfoot
Wife of Light
The Planets: A Cosmic Pastoral
One Hundred Names for Love
Dawn Light
The Moon by Whale Light

For children
Animal Sense
Monk Seal Hideaway
Bats: Shadows in the Night

Anthology
The Book of Love (with Jeanne Mackin)

The Zookeeper's Wife

Diane Ackerman

headline

The right of Diane Ackerman to be identified as the Author of
the Work has been asserted by her in accordance with the
Copyright, Designs and Patents Act 1988.

First published in paperback in 2013 by
HEADLINE PUBLISHING GROUP

1

Cataloguing in Publication Data is available from the British Library

Paperback ISBN 978 0 7553 6503 6

Typeset in Celeste by Palimpsest Book Production Limited,
Falkirk, Stirlingshire

Printed and bound in Great Britain by
Clays Ltd, St Ives plc

Headline's policy is to use papers that are natural, renewable and recyclable products and
made from wood grown in sustainable forests. The logging and manufacturing processes are
expected to conform to the environmental regulations of the country of origin.

Photos courtesy of the Warsaw Zoo

HEADLINE PUBLISHING GROUP
An Hachette UK Company
338 Euston Road
London NW1 3BH

www.headline.co.uk
www.hachette.co.uk

Contents

Author's Note

Jan and Antonina Żabiński were Christian zookeepers horrified by Nazi racism, who capitalized on the Nazis' obsession with rare animals in order to save over three hundred doomed people. Their story has fallen between the seams of history, as radically compassionate acts sometimes do. But in wartime Poland, when even handing a thirsty Jew a cup of water was punishable by death, their heroism stands out as all the more startling.

In telling their story, I've relied on many sources detailed in the bibliography, but most of all on the memoirs ("based on my diary and loose notes") of "the zookeeper's wife," Antonina Żabińska, rich with the sensuous spell of the zoo; her autobiographical children's books, such as *Life at the Zoo*; Jan Żabiński's books and recollections; and the interviews Antonina and Jan gave to Polish, Hebrew, and Yiddish newspapers. Whenever I say Antonina or Jan *thought, wondered, felt*, I'm quoting from their writings or interviews. I've also depended on family photographs (that's how I know Jan wore his watch on his hairy left wrist and Antonina had a thing for polka-dot dresses); conversations with their son Ryszard, various people at the Warsaw Zoo, and Warsaw women who were contemporaries of Antonina and also served the Underground; writings by Lutz Heck; artifacts viewed in museums, such as the dramatic Warsaw Uprising Museum and the eloquent Holocaust Museum in Washington, D.C.; the State Zoological Museum archives; the memoirs and letters collected by a secret group of wartime archivists who hid (in boxes and milk churns) documents that now reside at the Jewish Historical Institute in Warsaw; testimonies given to Israel's unique Righteous Among Nations program and the superb Shoah Project; and letters, diaries, sermons, memoirs, articles, and other writings by citizens of the Warsaw Ghetto. I studied how Nazism hoped, not only to dominate nations and ideologies, but to alter the world's ecosystems by extinguishing some countries' native species of plants and animals (including human beings), while going to great lengths to protect other endangered animals and habitats, and even to resurrect extinct species like the wild cow and forest bison. I pored over guides to Polish wildlife and plants (exploring Poland's natural world provided a steady stream of small astonishments); guides to Polish customs, cuisine, and folklore; and books on Nazi drugs, scientists, weapons, and other themes. I relished learning about Hasidism, Kabbalah, and pagan mysticism of the early twentieth century; Nazism's roots in the occult; and such practical matters as Polish social and political history and Baltic lampshades of the era.

I'm also indebted to the knowledge of my invaluable Polish advisor, Magda Day, who spent her first twenty-six years in Warsaw, and her daughter, Agata M. Okulicz-Kozaryn. On a trip to Poland, I gathered impressions at Białowieża Forest and in the Warsaw Zoo itself, where I loafed and prowled around the old villa and retraced Antonina's steps on the surrounding streets. I'm especially grateful to Dr. Maciej Rembiszewski, the current director of the Warsaw Zoo, and his wife, Ewa Zabonikowska, for their generosity of time and spirit, and also to the zoo staff for their knowledge, resources, and welcome. My thanks also go to Elizabeth Butler for her tireless and always upbeat assistance, and to Professor Robert Jan van Pelt for his careful critique.

I came to this story, as to all of my books, by a very personal route: Both of my maternal grandparents came from Poland. I've been intimately influenced by accounts of Polish daily life from my grandfather, who grew up in Letnia, a suburb of Przemyś, and left before World War II, and from my mother, some of whose relatives and friends lived in hiding or in the camps. My grandfather, who lived on a small farm, shared folk stories passed down through generations.

One of them tells of a village with a little circus whose lion had suddenly died. The circus director asked a poor old Jewish man if he would pretend to be the lion, and the man agreed since he needed the money. The director said: "All you have to do is wear the lion's fur and sit in the cage, and people will believe you're a lion." And so the man did, muttering to himself, "What strange jobs I've had in my life," when his thoughts were interrupted by a noise. He turned just in time to see another lion creeping into his cage and fixing him with a hungry stare. Trembling, cowering, not knowing how to save himself, the man did the only thing he could think of—vociferously chant a Hebrew prayer. No sooner had he uttered the first desperate words, *Shema Yisroel*

(Hear O Israel) . . . than the other lion joined in with *adonai elohenu* (the Lord our God), and the two would-be lions finished the prayer together. I could not have imagined how oddly relevant that folk story would be to this historical one.

The Zookeeper's Wife

Chapter 1

Summer, *1935*

At dawn in an outlying district of Warsaw, sunlight swarmed around the trunks of blooming linden trees and crept up the white walls of a 1930s stucco and glass villa where the zoo director and his wife slept in a bed crafted from white birch, a pale wood used in canoes, tongue depressors, and Windsor chairs. On their left, two tall windows crowned a sill wide enough for sitting, with a small radiator tucked below. Oriental rugs warmed the parquet floor, where strips of wood slanted together as repeating feathers, and a birch armchair squared one corner of the room.

When a breeze lifted the voile curtain enough for grainy light to spill in without casting shadows, barely visible objects began anchoring Antonina to the sensible world. Soon the gibbons would start whooping, and after that a pandemonium rip no one could sleep through, not owl-eyed student or newborn. Certainly not the zookeeper's wife. All the usual domestic chores awaited her each day, and she was clever with food, paintbrush, or needle. But she also had zoo problems of her own to solve, sometimes uncanny ones (such as hyena-cub soothing) that challenged her schooling and native gifts.

Her husband, Jan Żabiński, usually arose earlier, dressed in trousers and a long-sleeved shirt, and slid a large watch over his hairy left wrist before padding downstairs. Tall and slender, with a strong nose, dark eyes, and the muscular shoulders of a laborer, he was built a little like her father, Antoni Erdman, a Polish railroad engineer based in St. Petersburg, who traveled throughout Russia following his trade. Like Jan, Antonina's father had plenty of mental muscle, enough to get him and her stepmother, shot as members of the intelligentsia in the early days of the Russian Revolution in 1917, when Antonina was only nine. And like her father, Jan was a kind of engineer, though the connections he fostered were between people and animals, and also between people and their animal nature.

Balding, with a crown of dark brown hair, Jan needed a hat to fight burn in summer and chill in winter, which is why in outdoor photographs he's usually wearing a fedora, giving him an air of sober purpose. Some indoor photographs capture him at his desk or in a radio studio, jaw tight in concentration, looking like a man easily piqued. Even when he was clean-shaven, a five o'clock shadow stippled his face, especially on the philtrum between nose and mouth. A full, neatly edged upper lip displayed the perfect peaks women create with lip liner, a "Cupid's bow" mouth; it was his only feminine feature.

After the death of Antonina's parents, her aunt sent her to school full-time to study piano at the city's conservatory and also attend school in Tashkent, Uzbekistan, from which she graduated at fifteen. Before the year was out, they moved to Warsaw and Antonina took classes in foreign languages, drawing, and painting. She did a little teaching, passed an archivist's exam, and worked in the labeled past of Warsaw's College of Agriculture, where she met Jan, a zoologist eleven years her senior, who had studied drawing and painting at the Academy of Fine Arts, and shared her relish both for animals and animalistic art. When the position of zoo director came free in 1929 (the founding director had died after two years), Jan and Antonina leapt at the chance to shape a new zoo and spend their lives among animals. In 1931, they married and moved across the river to Praga, a tough industrial district with its own street slang, on the wrong side of the tracks, but only fifteen minutes by trolley from downtown.

In the past, zoos were privately owned and conferred status. Anyone could stock a curiosity cabinet, but it took means, and a little madness, to collect the largest crocodile, oldest turtle, heaviest rhino, rarest eagle. In the seventeenth century, King Jan III Sobieski kept many exotic animals at court; and rich noblemen sometimes lodged private menageries on their estates as a sign of wealth.

For years, Polish scientists dreamt of a big zoo in the capital to rival any in Europe, especially those in Germany, whose majestic zoos were famous worldwide. Polish children clamored for a zoo, too. Europe enjoyed a heritage of fairy tales alive with talking animals—some almost real, others deliciously bogus—to spark a child's fantasies and gallop grown-ups to the cherished haunts of childhood. It pleased Antonina that her zoo offered an orient of fabled creatures, where book pages sprang alive and people could parley with ferocious animals. Few would ever see wild penguins sledding downhill to sea on their bellies, or tree porcupines in the

5

Canadian Rockies, balled up like giant pinecones, and she believed that meeting them at the zoo widened a visitor's view of nature, personalized it, gave it habits and names. Here lived the *wild*, that fierce beautiful monster, caged and befriended.

Each morning, when zoo dawn arrived, a starling gushed a medley of stolen songs, distant wrens cranked up a few arpeggios, and cuckoos called monotonously like clocks stuck on the hour. Suddenly the gibbons began whooping bugle calls so crazy loud that the wolves and hunting dogs started howling, the hyenas gibbering, the lions roaring, the ravens croaking, the peacocks screeching, the rhino snorting, the foxes yelping, the hippos braying. Next the gibbons shifted into duets, with the males adding soft squealing sounds between their whoops and the females bellowing streams of long notes in their "great call." The zoo hosted several mated pairs, and gibbon couples yodel formal songs complete with overture, codas, interludes, duets, and solos.

Antonina and Jan had learned to live on seasonal time, not mere chronicity. Like most humans, they did abide by clocks, but their routine was never quite routine, made up as it was of compatible realities, one attuned to animals, the other to humans. When timelines clashed, Jan returned home late, and Antonina woke in the night to help midwife an animal like a giraffe (always tricky because the mother gives birth standing up, the calf falls headfirst, and the mother doesn't want help anyway). This brought a slated novelty to each day, and though the problems might be taxing, it imprinted her life with small welcome moments of surprise.

A glass door in Antonina's bedroom opened onto a wide

second-story terrace at the back of the house, accessible from each of three bedrooms and a narrow storage room they called the attic. Standing on the terrace, she could peer into the spires of evergreens, and over lilacs planted near six tall living room windows to catch river breezes and waft scent indoors. On warm spring days, the lilacs' purple cones swung like censers and a sweet narcotic amber drifted in at intervals, allowing the nose to rest awhile between fragrant reveilles. Perched on that terrace, inhaling air at the level of ginkgo and spruce, one becomes a creature of the canopies. At dawn, a thousand moist prisms ornament the juniper as one glances over the heavily laden limbs of an oak tree, beyond the Pheasant House, down to the zoo's main gate about fifty yards away on Ratuszowa Street. Cross over and you enter Praski Park, as many Warsawians did on warm days, when the linden trees' creamy yellow tassels drugged the air with the numbing scent of honey and the rhumba of bees.

Traditionally, lindens capture the spirit of summer—*lipa* means linden, and *Lipiec* means July. Once sacred to the goddess of love, they became Mary's refuge when Christianity arrived, and at roadside shrines, under lindens, travelers still pray to her for good fortune. In Warsaw, lindens enliven parks and ring cemeteries and markets; rows of tall, leaf-helmeted lindens flank the boulevards. Revered as God's servants, the bees they lure provide mead and honey for the table and beeswax candles for church services, which is why many churches planted linden trees in their courtyards. The bee-church connection became so strong that once, at the turn of the fifteenth century, the villagers of Mazowsze passed a law condemning honey thieves and hive vandals to death.

In Antonina's day, Poles felt less violent but still zealous about bees, and Jan kept a few hives at the far edge of the zoo, clustered like tribal huts. Housewives used the honey to sweeten iced coffee, make *krupnik*, hot vodka with honey, and bake *piernik*, a semisweet honey-spice cake, or *pierniczki*, honey-spice cookies. They drank linden tea to brighten a cold or tame the nerves. In this season, whenever Antonina crossed the park on her way to the trolley stop, church, or market, she walked through corridors thickly scented by linden flowers and abuzz with half-truths—in local slang, *lipa* also meant white lies.

Across the river, the skyline of Old Town rose from the early-morning mist like sentences written in invisible ink—first just the roofs, whose curved terra-cotta tiles over-lapped like pigeon feathers—then a story of sea-green, pink, yellow, red, copper, and beige row houses that lined cobble-stone streets leading to Market Square. In the 1930s, an open-air market served the Praga district, too, near the vodka factory on Ząbkowska (Tooth) Street designed to look like a squat castle. But it wasn't as festive as Old Town's, where dozens of vendors sold produce, crafts, and food below yellow and tan awnings, the shopwindows displayed Baltic amber, and for a few groschen a trained parrot would pick your fortune from a small jug of paper scrolls.

Just beyond Old Town lay the large Jewish Quarter, full of mazy streets, women wearing wigs and men sideburn curls, religious dancing, a mix of dialects and aromas, tiny shops, dyed silks, and flat-roofed buildings where iron balconies, painted black or moss green, rose one above the other, like opera boxes filled not with people but with tomato pots and flowers. There one could also find a special kind of *pierogi*, large chewy *kreplach*: fist-sized dumplings filled with seasoned stew meat and onions before being boiled, baked, then fried, the last step glazing and toughening them like bagels.

The heartbeat of eastern European Jewish culture, the Quarter offered Jewish theater and film, newspapers and magazines, artists and publishing houses, political movements, sports and literary clubs. For centuries, Poland had granted asylum to Jews fleeing persecution in England, France, Germany, and Spain. Some twelfth-century Polish coins even bear Hebrew inscriptions, and one legend has it that Jews found Poland attractive because the country's name sounded like the Hebrew imperative *po lin* ("rest here"). Yet anti-Semitism still percolated in twentieth-century Warsaw, a city of 1.3 million people, a third of whom were Jewish. They mainly settled in the Quarter, but also lived in posher neighborhoods throughout the city, though for the most part they kept their distinctive garb, language, and culture, with some speaking no Polish at all.

On a typical summer morning, Antonina leaned on the wide flat ledge of the terrace wall, where apricot tiles, cold enough to collect dew, dampened the sleeves of her red robe. Not all the bellowing, wailing, braying, and rumbling around her originated outside—some issued from the subterranean bowels of the villa, others from its porch, terrace, or attic. The Żabińskis shared their home with orphaned newborn or sick animals, as well as pets, and the feeding and schooling of lodgers fell to Antonina, whose animal wards clamored to be fed.

Not even the villa's living room was off limits to the animals. With its six tall window panels that could easily be

9

mistaken for landscape paintings, the long, narrow salon blurred the boundaries between inside and out. Across the room, a large wooden credenza displayed books, periodicals, nests, feathers, small skulls, eggs, horns, and other artifacts on its many shelves. A piano stood on an Oriental rug beside a scatter of boxy armchairs with red fabric cushions. In the warmest corner, at the far end of the room, dark brown tiles adorned a fireplace and hearth, and the sun-bleached skull of a bison rested atop the mantelpiece. Armchairs sat beside the windows, where afternoon light washed in.

One journalist who visited the villa to interview Jan was surprised by two cats entering the living room, the first with a bandaged paw and the second a bandaged tail, followed by a parrot wearing a metal neck cone, and then a limping raven with a broken wing. The villa bustled with animals, which Jan explained simply: "It's not enough to do research from a distance. It's by living beside animals that you learn their behavior and psychology." On Jan's daily rounds of the zoo by bicycle, a large elk named Adam swayed close behind, an inseparable companion.

There was something alchemical about living so intimately with the likes of lion kitten, wolf cub, monkey toddler, and eagle chick, as the animal smells, scratchings, and calls mingled with human body and cooking smells, with human chatter and laughter in a mixed family of den-mates. At first a new member of the household slept or fed on its old schedule, but gradually the animals began to live in synchrony as their rhythms drew closer together. Not their breathing, though, and at night the sleepy tempo of breaths and snufflings created a zoological cantata hard to score.

Antonina identified with animals, fascinated by how their senses tested the world. She and Jan soon learned to slow around predators like wild cats, because close-set eyes give them pinpoint depth perception, and they tend to get excited by quick movements a leap or two away. Prey animals like horses and deer enjoy wraparound vision (to spot predators creeping up on them), but panic easily. The lame speckled eagle, tethered in their basement, was essentially a pair of binoculars with wings. The hyena pups would have spotted Antonina coming in total darkness. Other animals could sense her approach, taste her scent, hear the faintest swoosh of her robe, feel the weight of her footsteps vibrating the floorboards a whisker's worth, even detect the motes of air she pushed aside. She envied their array of ancient, finely tuned senses; a human gifted with those ordinary talents, Westerners would call a sorcerer.

Antonina loved to slip out of her human skin for a while and spy on the world through each animal's eyes, and she often wrote from that outlook, in which she intuited their concerns and know-how, including what they might be seeing, feeling, fearing, sensing, remembering. When she entered their ken, a transmigration of sensibility occurred, and like the lynx kittens she hand-raised, she could peer up at a world of loud dangling beings:

> . . . with legs little or large, walking in soft slippers or solid shoes, quiet or loud, with the mild smell of fabric or the strong smell of shoe polish. The soft fabric slippers moved quietly and gently, they didn't hit the furniture and it was safe to be around them . . . calling "Ki-chi,

ki-chi," [until] a head with fluffy blond hair would appear and a pair of eyes behind large glass lenses would bend over. . . . It didn't take long to realize that the soft fabric slippers, the blond fluffy head, and the high-pitched voice were all the same object.

Often dabbling in such slippages of self, aligning her senses with theirs, she tended her wards with affectionate curiosity, and something about that attunement put them at ease. Her uncanny ability to calm unruly animals earned her the respect of both the keepers and her husband, who, though he believed science could explain it, found her gift nonetheless strange and mysterious. Jan, a devout scientist, credited Antonina with the "metaphysical waves" of a nearly shamanistic empathy when it came to animals: "She's so sensitive, she's almost able to read their minds. . . . She *becomes* them. . . . She has a precise and very special gift, a way of observing and understanding animals that's rare, a sixth sense. . . . It's been this way since she was little."

In the kitchen each morning, she poured herself a cup of black tea and started sterilizing glass baby bottles and rubber nipples for the household's youngest. As zoo nurse, she was lucky enough to adopt two baby lynxes from Białowieża, the only primeval forest left in all of Europe, an ecosystem Poles called a *puszcza*, a word evoking ancient woodlands undefiled by human hands.

Straddling what is now the border between Belarus and Poland, Białowieża unites the two at the level of antler and myth, and traditionally served both countries as a famous hunting retreat for kings and tsars (who kept an ornate lodge there), which, by Antonina's time, fell under the purview of scientists, politicians, and poachers. The largest land animals in Europe, European (or "forest") bison, sparred in its woods, and their decline helped to kindle Poland's conservation movement. As a bilingual Pole born in Russia who returned to Poland, she felt at home in that green isthmus linking different regimes, walking in the shade of trees half a millennium old, where the forest closes in, intimate as a tick, one fragile, fully furnished organism with no visible borders. Pristine acres of virgin forest, declared untouchable, create a realm that airplanes overfly by miles lest they scare the animals or taint the foliage. Looking up through the open parachutes of treetops, a visitor might spy a distant plane banking like a small silent bird.

Though outlawed, hunting still existed, leaving motherless young animals, the rarest of which usually arrived at the zoo in a crate marked "live animal." The zoo served as lifeboat, and during April, May, and June, the birthing season, Antonina expected crotchety offspring, each with its own special diet and customs. The month-old wolf cub would normally be tended by its mother and family members until two years old. The clean, sociable baby badger responded well to long walks and dined on insects and herbs. Striped wild boar piglets did justice to any table scraps. A red deer fawn bottle-fed until midwinter and skidded, splay-legged, on wooden floors.

Her favorites were Tofi and Tufa, the three-week-old lynx kittens, who needed bottle-nursing for six months and weren't really self-reliant for a year or so (and, even then, they liked walks on a leash down Praga's busiest street, while passersby gaped). Because so few wild lynxes remained in Europe, Jan went to Białowieża himself to fetch the kittens, and Antonina offered to raise them inside the house. When his taxi arrived at the main gate one summer evening, a guard ran to help Jan unload a small wooden box and together they carried it to the villa, where Antonina eagerly waited with sterilized glass bottles, rubber nipples, and warm formula. As they lifted off the lid, two tiny speckled fur balls stared up angrily at the human faces, hissed, and began biting and scratching any hand that reached for them.

"Human hands with so many moving fingers scare them," Antonina advised softly. "And our loud voices, and the sharp light from the lamp."

The kittens trembled, "half dead with fear," she wrote in her diary. Gently, she grabbed the scruff of one's neck, loose and hot, and as she lifted it from the straw, it hung limp and quiet, so she picked up the other one.

"They like it. Their skin remembers their mother's jaws carrying them from one place to another."

When she set them down on the floor in the dining room, they skittered around, exploring the slippery new landscape for a few minutes, then hid under a wardrobe as if it were a rock overhang, inching way back into the darkest crevices they could find.

In 1932, abiding by Polish Catholic tradition, Antonina chose a saint's name for her own newborn son, Ryszard, or Ryś for short—the Polish word for lynx. Though not part of the zoo's "four-legged, fluffy, or winged" brigade, her son joined the household as one more frisky cub that babbled and clung like a monkey, crawled around on all fours like a bear, grew whiter in winter and darker in summer like a wolf. One of her children's books describes three household toddlers learning to walk at the same time: son, lion, and chimpanzee. Finding all young mammals adorable, from rhino to possum, she reigned as a mammal mother herself and protectress of many others. Not an outlandish image in a city whose age-old symbol was half woman, half animal: a mermaid brandishing a sword. As she said, the zoo quickly became her "green kingdom of animals on the right side of the Vistula River," a noisy Eden flanked by cityscape and park.

Chapter 2

"Adolf has to be stopped," one of the keepers insisted. Jan knew he didn't mean Hitler but "Adolf the Kidnapper," a nickname given to the ringleader of the rhesus monkeys, who had been waging war with the oldest female, Marta, whose son Adolf had stolen and given to his favorite mate, Nelly, who already had one baby of her own. "It's not right. Each mother should feed her own baby, and why deprive Marta of her baby just to give Nelly two?"

Other keepers offered health bulletins about the zoo's best-known animals, like Rose the giraffe, Mary the African hunting dog, Sahib the petting-zoo colt, who had been sneaking into the pasture with the skittish Przywalski horses. Elephants sometimes develop herpes on their trunks, and in captive settings, an avian retrovirus or an illness like tuberculosis passes easily from humans to parrots, elephants, cheetahs, and other animals, and back again to humans—especially in Jan's preantibiotic era, when serious infection could savage a population, animal or human. That meant calling the zoo vet, Dr. Lopatynski, who always arrived on his spluttering motorcycle wearing a leather jacket, big hat with long waving earflaps, cheeks whisked red by the wind, and pince-nez glasses perched on his nose.

What else might have been discussed at the daily meetings? In an old zoo photograph, Jan stands beside a large half-excavated hippo enclosure that's partly braced with heavy wooden ribs, the sort that flex ship hulls. The background vegetation suggests summer, and all digging had to be finished before the ground hardened, which can happen as early as October in Poland, so it's likely he demanded progress reports and chivvied the foreman. Thievery posed another worry, and since the exotic animal trade flourished, armed guards patrolled day and night.

Jan's grand vision of the zoo shines through his many books and broadcasts; he hoped that one day his zoo might achieve an illusion of native habitats, where natural enemies could share enclosures without conflict. For that mirage of a primal truce one needs to recruit acres of land, dig interlocking moats, and install creative plumbing. Jan planned an innovative zoo of world importance at the heart of Warsaw's life, both social and cultural, and at one point he even thought of adding an amusement park.

Basic concerns for zoos both antique and modern include keeping the animals healthy, sane, safe, and above all contained. Zoos have always faced ingenious escape artists, leggy lightning bolts like klipspringers, which can leap right over a man's head and land on a rock ledge the size of a quarter. Powerful and stocky with an arched back, these nervous little antelopes only weigh forty pounds, but they're agile and jump on the tips of their vertical hooves like ballet dancers performing on their toenails. Startle them and they will bounce around the enclosure and possibly leap the fence, and, like all antelopes, they *pronk*. Legend has it that, in 1919, a Burmese man invented the closest human equivalent to pronking—a hopping stick for his daughter, Pogo, to use crossing puddles on her way to school.

After the jaguar nearly cleared its moat at the current Warsaw Zoo, Dr. Rembiszewski planted an electric fence of the sort farmers use to jolt deer from their crops, only custom-built and much higher. Electric fences were available to Jan, who may well have priced one and discussed its feasibility given the layout of the big cats' enclosure.

After breakfast each day, Antonina walked to the zoo office building and awaited VIP visitors, because besides running the household and nursing sick animals, she greeted distinguished guests from Poland and abroad and welcomed press or government officials. Guiding people round, Antonina amused them with anecdotes and curiosities absorbed from books, Jan's talks, or observed firsthand. As they strolled through the zoo, they glimpsed versions of wetlands, deserts, woods, meadows, and steppes. Some areas stayed shaded, others swam in sunlight, and strategically arranged trees, shrubs, and rocks offered shelter from winter's hammering winds that could claw the roof off a barn.

She began at the main gate on Ratuszowa Street, facing a long straight boulevard flanked by enclosures where the first thing to catch a visitor's eye was a wobbly pink pond—pale flamingos strutting with backward-bent red knees, their mouths black change-purses. Not as vivid as wild flamingos, tinted coral pink from eating crustaceans, they were eye-catching enough to be the zoo's receptionists, and full of raucous growls, grunts, and honking. Just beyond them one met cages of birds from all over the world: noisy, colorfully plumed exotics like mynas, macaws, marabous, and crowned cranes; as well as native birds like the diminutive pygmy owl, or the giant eagle owl that can snatch up a rabbit in its talons.

Peacocks and small deer roamed the zoo as they pleased, trotting away when people approached, as if pushed by an invisible wave. Atop a small grassy mound, a female cheetah sunned herself while her speckled kittens leapt and wrestled nearby, occasionally distracted by the free-range deer and peacocks. Tantalizing as loose prey must have been for caged lions, hyenas, wolves, and other predators, it also kept their senses keen and added a carnal edge to their day. Black swans, pelicans, and other marsh and water birds floated on a dragon-shaped pond. To the left, open enclosures revealed grazing forest bison, antelopes, zebras, ostriches, camels, and rhinos. To the right, visitors viewed tigers, lions, and hippos. Then, following the gravel path, they circled back past the giraffes, reptiles, elephants, monkeys, seals, and bears. The villa lay nearly hidden among the trees, within hooting distance of the aviaries, just before one got to the chimps, due east of the penguins.

The grasslands habitats included African wild dogs, excitable long-legged canines always on the run, swinging their wide heads and sniffing suspiciously as they swiveled large stiff ears. Their scientific name, *Canis pictus* (painted dog), suggests the beauty of their fur, randomly splotched with yellow, black, and red. But not their ferocity or endurance: they could drag down a bolting zebra or chase an antelope for miles. The zoo boasted the first in Europe, a real prize, even if in Africa farmers regarded them as vicious pests. In Warsaw they were picturesque showmen, no two patterned the same, and a crowd always formed in front of them. The zoo also bred the first Grewyi zebras, native to Abyssinia, which look familiar at first until you realize that, unlike textbook zebras, they're taller and more heavily striped, with narrow bars that converge vertically around the body and run horizontally down the legs, striping all the way to the hooves.

And then there was Tuzinka, still covered in baby fuzz, one of only twelve elephants ever born in captivity. Hence her name, from *tuzin*, the Polish word for a dozen. Antonina had midwifed Kasia when she gave birth to Tuzinka, at 3:30 A.M. on a cool April morning. In her diary she described Tuzinka as a giant bundle, the largest baby animal she'd ever seen, weighing in at 242 pounds, standing a little over three feet tall, with blue eyes, a down of black hair, large pansy-like ears, a tail that seemed too long for her body—a wobbly confused newborn dropping into life's sensory bazaar. Her blue eyes flickered with the same surprise Antonina beheld in the eyes of other newborn animals—gawking, fascinated, yet baffled by all the shine and clangor.

To nurse, Tuzinka stood beneath her mother, back knees bent, reaching up with her soft mouth. The look in her eyes signaled that nothing existed but the flow of warm milk and the drum of her mother's reassuring heartbeat. That's how photographers captured her, in 1937, for a black-and-white postcard that proved a popular souvenir, as did a stuffed cloth baby elephant. Old photographs show delighted visitors reaching out to Tuzinka and her mother, who is reaching back with an extended trunk, across a small moat edged with short metal spikes. Since elephants don't jump, a six-foot-deep trench that's six feet wide at the top and narrower at the bottom will trap them, provided the elephants don't fill the trench in with mud and wade across, as some have been known to do.

Animal smells created the zoo's olfactory landscape, some subtle, some almost sickening at first. Especially the scent signposts of hyenas, which turn their anal pouches inside out and ooze a stinky paste known in the trade as "hyena butter." Each foul-smelling ad lasts a month or so, broadcasting news, and a mature male paints about a hundred fifty a year. Then there's the hippo's dominance display of defecating while propelling its little tail, flinging dung everywhere. Male musk oxen habitually sprinkle themselves with their own urine, and because sea lions trap rotting food between the teeth, their breath reeks a yard away. The kakapo, a black-feathered flightless parrot with a shocking white eye and orange beak, smells like an old clarinet case. During mating season, male elephants dribble a powerful sweet musth from a little gland near each eye. The crested auklet's feathers smell of tangerine, especially during breeding season, when courting auklets poke their beaks into each other's pungent neck-ruff. All the animals telegraph scent codes as distinctive as calls, and after a while, Antonina grew used to the thick aroma of their agendas—biological threats, come-ons, and news reports.

Antonina felt convinced that people needed to connect more with their animal nature, but also that animals "long for human company, reach out for human attention," with a yearning that's somehow reciprocal. Her imaginary transits into the *Umwelt* of animals banished the human world for a spell, a realm of saber-rattling and strife where parents suddenly vanish. Playing chase and tumble games with the lynx kittens, feeding them by hand, releasing herself to the sandy lick of warm tongues on her fingers and the insistent kneading of paws, as the no-man's-land between tame and wild softened even more, helped her forge a bond with the zoo she described as "everlasting."

The zoo also offered Antonina a pulpit for conservation, a sort of walking ministry, evangelism beside the Vistula as a tour of lesser gods, and she offered visitors a unique bridge to nature. But first they had to cross the cagelike bridge spanning the river and enter the woolier side of town. When she told them absorbing stories about lynxes and other animals, the earth's vast green blur reeled into focus briefly as a single face or motive, a named being. She and Jan also encouraged directors to stage film, music, and theater events at the zoo, and loaned animals for roles in shows when asked—lion cubs being the most popular. "Our zoo was full of life," she wrote. "We had lots of visitors: young people, animal lovers, and just visitors. We had many partners: universities in Poland and abroad, the Polish Health Department, and even the Academy of Fine Arts." Local artists crafted the zoo's stylized Art Deco posters, and the Żabińskis invited artists of all stripes to come and uncage their imaginations.

Chapter 3

One day on his bicycle rounds of the zoo, Jan left Adam the elk to graze on the lawn and shrubbery and entered the warm birdhouse, redolent with moist hay and lime. There, a petite woman stood close to a cage, moving her elbows in mimicry of the birds as they preened and posed. With her dark wavy hair, compact body, and thin legs sticking out from the hem of her smock, she almost qualified for enclosure herself. Bobbing on a trapeze overhead, a walleyed parrot screeched: *"What's your name? What's your name?"* And in a melodious voice, the woman piped back: *"What's your name? What's your name?"* The parrot leaned down and eyed her hard, then turned its head and fixed her with the other eye.

"Good day," Jan said. *Dzień dobry.* It's the way Poles began most polite verbal exchanges. She introduced herself as Magdalena Gross, a name Jan knew well, since Gross's sculptures were commissioned by wealthy Poles and international admirers alike. He didn't know she sculpted animals, but then neither did she before that day. Later she'd tell Antonina that when she first visited the zoo she'd been so captivated that her hands had started molding air, so she decided to bring her tools and go on safari, and fate led her to that enclosure with birds streamlined like futuristic trains. Jan kissed her hand lightly according to Polish custom, said it would honor him if she considered the zoo her open-air studio and the animals her fidgety models.

By all accounts, tall, slender, fair Antonina looked like a Valkyrie at rest, and short, dark, Jewish Magdalena vibrated with energy. Antonina saw Magdalena as a winning array of contradictions: emphatic yet vulnerable, daring but modest, zany yet highly disciplined, someone excited by life—which may be what appealed most to Antonina, who wasn't as stoic or solemn as Jan. The two women shared a passion for art and music, as well as a similar sense of humor, were close in age, and had friends in common—thus began what would become an important friendship. What would Antonina have served when Magdalena joined her for tea? Most Warsawians offer black tea and sweets to guests, and Antonina raised roses and jarred a lot of preserves, so at some point she's bound to have prepared the traditional Polish pastry of soft doughnuts filled with a layer of pink rose-petal jam and coated in an orange glaze that smells of fire.

Magdalena confided that she'd been feeling stale and uninspired, her creative attic empty, when she happened by the zoo one day and saw a shocking flock of flamingos strutting by. Beyond them roamed a dream-panoply of even stranger animals—fabulous shapes, and hues more subtle than any painter could mix. The spectacle hit her with all the power of revelation, and inspired a series of animal sculptures that would win international acclaim.

The zoo looked magnificent by the summer of 1939, and Antonina began making elaborate plans for the following spring, when she and Jan would have the honor of hosting the International Association of Zoo Directors' annual meeting in Warsaw. However, that meant pushing to the rim of awareness such seismic fears as: *if our world's still intact.* Nearly a year earlier, in September 1938, when Hitler had seized Sudetenland, a part of Czechoslovakia bordering Germany and populated mainly by Germans, France and Britain had acquiesced, but Poles worried about their own borderlands. German territory ceded to Poland during 1918–22 included eastern Silesia and the region formerly known as the Pomeranian Corridor, an act that effectively separated East Prussia from the rest of Germany. The important German Baltic port of Gdańsk had been declared a "Free City," open to both Germans and Poles.

A month after he invaded Czechoslovakia, Hitler demanded the return of Gdańsk and the right to build an extraterritorial road through the Corridor. Diplomatic wrangling in early 1939 led to antagonism by March, when Hitler secretly ordered his generals to "deal with the Polish question." Relations between Poland and Germany gradually disintegrated and Poles awoke to omens of war, a horrifying thought but not a new one. Germany had occupied Poland so often since the Middle Ages, most recently in 1915–18, that Slavs fighting Teutons had achieved the status of patriotic tradition. Cursed by its strategic location in eastern Europe, Poland had been invaded, sacked, and carved up many times, its borders ebbing and flowing; some village children learned five languages just to speak with neighbors. War wasn't something Antonina wanted to think about, especially since her last experience of war stole both of her parents, so she assured herself, as most Poles did, of their solid alliance with France, keeper of a powerful army, and Britain's sworn protection. Optimistic by nature, she concentrated on her fortunate life. After all, in 1939, not many Polish women could be thankful for a good marriage, a healthy son, and a satisfying career, let alone a wealth of animals she regarded as stepchildren. Feeling blessed and high-spirited, Antonina took Ryś, his aged nanny, and Zośka the St. Bernard to the small popular vacation village of Rejentówka in early August, while Jan stayed in Warsaw to oversee the zoo. She also decided to include Koko, an elderly pink cockatoo prone to dizziness who often fell off her perch. Because Koko had a nervous habit of plucking out her breast feathers, Antonina dressed her in a metal collar that acted like a megaphone for squawks, and hoped "the fresh forest air, getting to eat wild roots and twigs," might cure her ills and return her colorful plumage. The by now full-grown lynxes stayed behind, but she carted along a new arrival, a baby badger named Borsunio (Little Badger), too young to leave unattended. Most of all, she wanted to spirit Ryś away from Warsaw, rife with talk of war, for one last summer of innocent play in the countryside, his and hers both.

The Żabińskis' country cottage nestled in a forest hollow four miles from a wide gall in the Bug River, and only minutes from its small tributary, the Rządza. Antonina and Ryś arrived on a hot summer day, with the smell of pine resin in the air and waves of acacias and petunias in bloom, the last rays of sunlight lighting the tips of old trees and darkness already fallen in the lower reaches of the forest, where the shrill rib-music of cicadas mixed with the descending calls of cuckoos and the whine of hungry female mosquitos.

A moment later, on one of the small verandas, she could submerge in the shadow of a scented grapevine "smelling of its faint, hardly perceptible flowers, but nicer than a rose, than lilac and jasmine, than the sweetest smell—golden lupin from the field," while "only a few steps through the overgrown grass stood . . . the forest wall, towering, young-green from the oak, slashed with white birch here and there. . . ." She and Ryś sank into the green quiet that seemed light-years from Warsaw, a huge, internal, personal distance, not just miles. Without even a radio in the cottage, nature provided the lessons, news, and games. One popular local pastime involved going into the forest and counting the aspens.

Each summer the cottage awaited them with dishes, pots, a washtub, sheets, and a large surplus of dry provisions, and they provided the ensemble of human and animal characters that transformed it from bungalow to burlesque. After they settled the large birdcage stand on the veranda and fed the cockatoo bits of orange, Ryś attached a halter to the badger and tried to persuade it to walk on a leash, which it did, but only in reverse, pulling Ryś along at speed. Like the other animals in her circle, Badger warmed to Antonina, who referred to him as her "foster child" and taught him to come to his name, paddle with them in the river, and climb up onto her bed to nurse from a bottle. Badger taught himself to scratch at the front door to be let out for toileting, and he bathed sitting back in a washtub, human style, while splashing sudsy water against his chest with both arms. In her diary, she noted how Badger's instincts mixed with human customs and his own one-of-a-kind personality. Scrupulous about toileting, for instance, he dug one toilet hole on each side of the house and galloped home from long walks just to use them. But one day, when she couldn't find Badger, she checked all of his usual daytime napping places—a drawer in the linen closet, between her bed's sheet and duvet cover, in Ryś's nanny's suitcase—with no success. In Ryś's bedroom, she stooped to look under the bed and glimpsed Badger pushing Ryś's training potty out into the open, climbing onto the white enamel bowl, and using it as it was intended.

Near the end of summer vacation, Ryś's friends Marek and Zbyszek (sons of a doctor who lived on the other side of Praski Park) stopped by on their way home from Hel Peninsula on the Baltic Sea, full of chatter about the many ships moored in Gdynia harbor, the smoked fish and sailing trips, and all the changes to the waterfront. Sitting in the dimly lit living room while night slipped around her, Antonina overheard the boys on the porch steps talking about their summer adventures, and she realized that for Ryś the Baltic Sea he'd visited three years earlier probably existed only as a hazy memory that included the crashing surf and the glassy heat of noon sand.

"You won't believe how they've dug up the beach! Next year there won't be a civilian on it," Marek said.

"But why?" Ryś asked.

"To build fortifications, for war!"

His older brother eyed him hard, and Marek wrapped an arm around Ryś's shoulder and said dismissively: "But who cares about the beach. Tell us about Badger instead."

And so Ryś began, stuttering a little at first, then developing animation and speed, to tell of forest pirates and the shenanigans of Badger, culminating in the night Badger spilled a bedside bucket of cold water onto a sleeping neighbor lady he'd crawled into bed with, and the boys fell together, twitching with laughter.

"It feels good to hear them laughing," Antonina thought, "but this perpetual splinter poking into Ryś—the war—it's still a murky idea to him. He only associates words like *torpedo* and *fortifications* with toys, the beautiful ships he floats in bays surrounding the sand forts he builds along the shore of the Rządza. And there's his enchanting game of cowboys and Indians when he shoots pinecones with bow and arrow . . . but the other possibility, of a *real* war, that he doesn't understand yet, thank heavens."

The older boys believed, as Antonina did, that war belonged to the world of adults, not children. She sensed that Ryś yearned to grill them with questions, though he didn't want to look stupid or, worse, like a little kid, so he kept quiet about the invisible hand grenade lying at his feet that everyone feared might explode.

"What a subject to be broached by the innocent lips of children," Antonina reflected, glancing at the sun-bronzed faces of the three boys, glistening in the light cast by a large oil lamp. "Grazed by sadness" about their safety, she wondered yet again: "What will become of them, if war begins?" It was a question she'd been denying, sidestepping, and rewording for months. "Our animal republic," she finally admitted to herself, "exists in the busiest and most buzzing Polish city, as a small autonomous state defended by the capital. Living behind its gates, as if on an island cut off from the rest of the world, it seems impossible the waves of evil spilling across Europe could overwhelm our little island as well." As darkness began seeping into everything, erasing edges, a free-floating anxiety plagued her; eager as she was to mend the fabric of her son's life the instant holes formed, she could only await the unraveling.

She meant the last idylls of summer to be well spent, so the next morning she organized a mushroom-hunting brigade, with prizes and honors for whoever bagged the most saffron milk-cap, boletus, and button mushrooms, which she planned to jar. If war did erupt, spreading mushroom marinade on bread in winter would fill everyone with cottage memories of river swims, Badger's antics, and better days. They hiked four miles to the Bug River, with Ryś carried piggyback at times, Zośka trotting alongside, and Badger riding in a knapsack. Stopping by meadows along the way, they picnicked and played soccer, with Badger and Zośka serving as goalies, though Badger battled fiercely over the leather ball once he latched onto it with teeth and claws.

Most summer weekends, Antonina left Ryś with his nanny in the country and returned to Warsaw to spend a few days alone with Jan. On Thursday, August 24, 1939, the same day Britain renewed its pledge to aid Poland if Germany invaded, Antonina made her usual visit to Warsaw, where, to her shock, she saw antiaircraft sites ringing the city, civilians digging trenches and erecting barricades, and, most disturbing of all, posters that announced an imminent draft. Only the day before, foreign ministers Ribbentrop and Molotov had stunned the world by revealing that Germany and the Soviet Union had signed a nonaggression pact.

"The only thing dividing Berlin from Moscow is Poland," she thought.

Neither she nor Jan knew of the pact's secret clauses, already hacking up Poland after a two-stage invasion and divvying up its desirable farmlands.

"Diplomats are cagey. This might only be a bluff," she thought.

Jan knew Poland hadn't the planes, weapons, or war equipment to compete with Germany, and so they started talking seriously about sending Ryś somewhere safer, to a town of no military interest, if such a place existed.

Antonina felt as though she were "waking from a long dream, or entering a nightmare," either way a psychic earthquake. Vacationing far from the political clatter of Warsaw, cocooned in "the calm, level order of the farmer's life, the harmony of white sand dunes and weeping willows," with each day enlivened by eccentric animals and a little boy's adventures, it had been nearly possible to ignore world events, or at least stay optimistic about them, even stubbornly naïve.

Chapter 4
Warsaw, September 1, 1939

Just before dawn, Antonina woke to the distant sound of gravel pouring down a metal chute, which her brain soon deciphered as airplane engines. *Let them be Polish planes on maneuver,* she prayed as she went onto the terrace and scouted a strange sunless sky, veiled as she'd never seen it before, not with clouds but a thick golden-white gloss hanging low on the ground like a curtain, yet not smoke, not fog, and stretching along the horizon from rim to rim. A veteran of World War I and a reserve officer, Jan had spent the night on duty, but she didn't know where, just "somewhere outside the zoo," in the city canyons beyond the mental moat of the Vistula.

She heard "the hum of planes, tens, maybe even hundreds," that sounded like "faraway surf, not a calm surf but when waves crash onto the beach during a storm." Listening a moment longer, she detected the telltale unsynchronized hum of German bombers that Londoners, later in the war, swore they heard grumbling, *Where are you? Where are you? Where are you?*

Jan returned home by 8 A.M., agitated and with only sketchy information. "Those won't be the practice maneuvers we were told about," he said. "They'll be bombers, Luftwaffe squadrons escorting the approaching German army. We've got to leave right away." With Ryś and his nanny safe in Rejentówka, they decided to head first to the nearer village of Zalesie, where his cousins lived, but waited for further updates on the radio.

This was the first day of the school year for Polish children, a day when the sidewalks should have been streaming with school uniforms and knapsacks. From the terrace, they saw Polish soldiers sprinting in all directions—down the streets, over lawns, even into the zoo—erecting balloon barriers, aligning antiaircraft guns, and piling long black cannon shells tapered at one end like animal droppings.

The zoo animals seemed unaware of danger. Small fires didn't scare them—for years they'd trusted the sight of household bonfires—but they grew alarmed by the sudden flood of soldiers, because the only humans they'd ever seen in early morning were the dozen or so blue-uniformed keepers, usually with food. The lynxes began gargling a sound between roar and meow, the leopards chug-chugged low notes, the chimpanzees yipped, the bears brayed like donkeys, and the jaguar sounded as though it were hacking up something lodged in its throat.

By 9 A.M., they'd learned that, to justify invasion, Hitler had staged a phony attack on the German border town of Gleiwitz, where SS troops dressed in Polish uniforms commandeered a local radio station and broadcast a fake call to arms against Germany. Although foreign journalists imported to witness the events were shown dead bodies of prisoners (dressed in Polish uniforms) as proof of hostilities, no one fell for the charade. Still, even such a hoax couldn't go unanswered, and at 4 A.M. Germany's battleship *Schleswig-Holstein* bombarded an ammunition depot near Gdańsk, and Russia's Red Army started preparing to invade from the east.

Antonina and Jan packed hurriedly and set out on foot across the bridge, hoping to make their way to Zalesie, beyond the Vistula River and only a dozen miles to the southeast. As they approached Zbawiciel Square, the engine noise ground louder and then planes floated overhead, appearing in the gap between the rooftops like stereopticon slides. Bombs whistled down and struck a few streets in front of them, with black smoke followed by the crackling of shattered stucco roof tiles and the rasp of brick and mortar crumbling.

Every bomb creates a different scent, depending on where it hits, what it boils into aerosol and the nose detects slipping apart, as molecules mix with air and float free. Then the nose can pick up ten thousand distinctive scents, from cucumber to violin resin. When a bakery was hit, the rising dirt cloud smelled of yeast sours, eggs, molasses, and rye. The mingled odors of cloves, vinegar, and burning flesh spelled the butcher's. Charred flesh and pine meant an incendiary bomb that blasted houses with a hot, fast fire, and that the people inside had died quickly.

"We've got to turn back," Jan said, and they ran past the walls of Old Town and across the whistling metal bridge. At the zoo once more, Antonina noted: "I was so depressed that I couldn't do anything. I could only hear Jan's voice directing his staff: 'Bring a cart with horses, load it with food and coal, pack warm clothes, and go immediately. . . .'"

For Jan, the puzzle of finding a town of no military interest posed an equation littered with unknowns he wasn't prepared for, since neither he nor Antonina had thought the Germans would invade Poland. They'd worried, but agreed "it was only fear speaking": a private siege, not signs of an impending war. Antonina wondered how they could have guessed so wrong, and Jan concentrated on hiding his family somewhere safe while he remained at the zoo to tend the animals as long as possible and await orders.

"Warsaw will soon be closed," he reasoned, "and the German army is advancing from the east, so I think it's best if you return to the cottage in Rejentówka."

She pondered this, then decided, despite misgivings: "Yes, at least it's a place we know, one Ryś associates with good times." Really, she had no idea, but persisted in packing, relying on Jan's hunch, then climbed into a cart loaded for what might be a long absence, and set out quickly before the roads grew too crowded.

The resort village of Rejentówka lay only about twenty-five miles away, but Antonina and a cart driver spent seven hours en route, sharing the dirt road with thousands of people, mainly on foot, since cars, trucks, and most horses had been confiscated by the military. Women, children, and old men hurried along in a worried trance, escaping the city with whatever they could carry, some pushing baby buggies, wagons, and hand trolleys, some hauling suitcases and small children, but most wearing several layers of clothing, with knapsacks, bags, and shoes slung across their bodies or swinging from their necks.

Flanking the road, tall poplars, pine, and spruce juggled large brown balls of mistletoe in their limbs; and black-and-white storks nested atop the telephone poles, still fattening up for their arduous flight to Africa. Soon farm fields quilted both sides of the road, grain glistening and tassels pointing skyward. Antonina wrote of sweat pouring in rivulets and breath bunching, the air clotted with dust.

A storm's distant rumble became a cloud of gnats on the horizon, then grew to German aircraft looming near in seconds, chewing up the skies, flying low overhead, panicking people and horses alike. Pelted by bullets, everyone hurried through clouds of flying dirt, the unlucky fell, and the relatively lucky fled beneath splattering machine-gun fire. Dead storks, redwings, and rooks littered the road along with tree branches and dropped satchels. Catching a bullet was sheer chance and for seven hours Antonina beat the odds, but not without scenes of the dead and dying etched into memory.

At least her son, in Rejentówka, was spared these images, so hard to erase, especially for a small child whose brain, busily sampling the world, was learning what to expect and stitching those truths in place at a trillion connections. *Stay prepared for this world the rest of your life*, a child's brain tells itself, *a world of mayhem and uncertainty*. "That which doesn't kill you, makes you stronger," Nietzsche wrote in *The Twilight of the Idols*, as if the will could be annealed like a Samurai sword that is heated and pounded, bent and reforged, until it becomes indestructible. But the metal of a little boy, what does the pounding do to him? Mixed with Antonina's worry about her son was moral outrage that the Germans "in this modern war, so different from wars we knew, allowed the killing of women, children, and civilians."

As the dust settled, blue sky returned and she noticed two Polish fighters attacking a heavy German bomber above a field. From afar, the geometry of the scene looked homely, like fierce wrens driving off a hawk, and people cheered whenever the fighters stung the bomber with tufts of smoke. Surely an air force that agile could repel the Luftwaffe? Threads of tinsel flashed in the waning sunlight, and suddenly the bomber gushed a fountain of blood-red flames and fell to earth in a sharp curve. Then a white jellyfish floated above the peaks of the pine trees: a German pilot swaying under his parachute, slowly descending through a cornflower-blue sky.

Like many Poles, Antonina didn't realize the magnitude of danger, relying instead on a Polish air force that boasted superbly trained and famously courageous pilots (especially those of the Pursuit Brigade defending Warsaw), whose outnumbered, obsolete PZL P.11 fighters posed no match for Germany's fast, swervy Junkers JU-87 Stukas. Polish Karas bombers swooped low over German tanks at such a slow speed, while flying level, that they fell easy prey to antiaircraft fire. She didn't know that Germany was testing out a new form of combined-arms warfare which would come to be called *Blitzkrieg* (lightning war), a charge-in-with-everything-you've-got—tanks, planes, cavalry, artillery, infantry—to surprise and terrify the enemy.

When she finally arrived in Rejentówka, she found a ghost town with summer guests gone, shops shuttered for the season, and even the post office closed. Exhausted, rattled, and dirty, she rode to the cottage hemmed in by tall trees and luminous quiet, in a setting that smelled familiar and safe, full of the mingled aromas of loam, meadow herbs and wild grasses, decaying wood and pine oil. One can picture her hugging Ryś hard and greeting his nanny; eating a dinner of buckwheat, potatoes, and soup; unpacking; bathing; longing for the habitual routines of just another summer, but unable to calm her nerves or quell her sense of foreboding.

Over the next few days, they often stood on the porch watching waves of German planes, en route to Warsaw, blacken the sky in lines neat as hedgerows. The regularity addled her: each day planes swarmed above at 5 A.M. and again after sunset, without her knowing whom exactly they had bombed.

The local landscape looked strange, too, since Rejentówka wasn't a spot they visited in autumn, without vacationers and pets. Tall lindens had begun turning bronze and oaks the burnt maroon of stale blood, while some green survived on the maples, where yellow-bellied evening grosbeaks fed on winged seeds. Along the sandy roads, staghorn sumac shrubs raised antler-velvet twigs and cone-shaped clusters of hairy red fruits. Blue chicory, brown cat-o'-nine-tails, white dame's rocket, pink thistle, orange hawkweed, and goldenrod tuned the meadows to fall, in a tableau that changed whenever a breeze bent the stems like a hand gliding over a plush carpet.

On September 5, Jan arrived by train, his face somber, to find Antonina "very depressed and confused."

"I've heard rumors that a wing of the German army, invading from East Prussia, will soon reach Rejentówka," he told her. "But the front hasn't arrived in Warsaw yet, and people are slowly getting used to the air raids. Our army is bound to protect the capital at all costs, so we may as well return home."

Even if he didn't sound altogether convinced, Antonina

agreed, in part because Jan was a good strategist whose hunches usually panned out, but she also thought how much easier life would be if they could stay together, sharing comforts, worries, and fears. Traveling the main road again was out of the question.

At night, they boarded a slow train with blackened windows and arrived in civil morning twilight, the hour of brightening before the sun spills over the horizon, in a lull between the night and dawn raids. According to Antonina, horses awaited them at the station and they rode home bewitched by the everyday—windless calm, damp air, aster hedges, colorful leaves, squeaky axles, clopping hooves on cobblestone—and, for a short spell, they slipped into the premechanized past, sinking deep into a pristine stillness where the war seemed to her muffled and unreal, only a remote glow like the moon.

At the main gate in Praga, the toll smacked her wide awake again as she dismounted. Bombs had ripped up the asphalt, shells had bitten large chunks out of the wooden buildings, cannon wheels had furrowed the lawns, old willows and lindens dangled unplugged limbs. Antonina held Ryś tight, as if the desolation before her were communicable. Unfortunately, the zoo edged a river with busy bridges, prime German targets, and with a Polish battalion stationed there, it had made a superb target, repeatedly, over several days. Picking their way through debris, they walked to the villa and its bomb-cratered yard. Antonina's eyes fell to the flower beds crushed from the hooves of horses, and she fixed on the small delicate calyxes of flowers stomped into the ground "like colorful teardrops."

Just after dawn, the day and battle started heating up.

Standing on the front porch, they were surprised by the canyon echoing of hoarse explosions and snapping iron girders. Suddenly the ground trembled and walked under their feet, and they hurried indoors, only to find the roof beams, floors, and walls all shaking. The moaning of lions and yowling of tigers spiraled from the big cat house, where she knew cat mothers, "crazy with fear, were grabbing their young by the scruff of the neck and pacing their cages, anxiously looking for a safe place to hide them." The elephants trumpeted wildly, the hyenas sobbed in a frightened sort of giggle interrupted by hiccups, the African hunting dogs howled, and the rhesus monkeys, agitated beyond sanity, battled one another, their hysterical shrieks clawing the air. Despite the uproar, workers continued to carry water and food to the animals and check their cage bars and locks.

In this Luftwaffe attack, a half-ton bomb destroyed the polar bears' mountain, smashing the walls, moats, and barriers and freeing the terrified animals. When a platoon of Polish soldiers found the panicky bears, ribboned with blood and circling round their old haunt, they quickly shot them. Then, fearing lions, tigers, and other dangerous animals might escape, too, the soldiers decided to kill the most aggressive ones, including the male elephant, Jaś, Tuzinka's father.

Watching from the front porch, Antonina had a good view across the grounds to where Polish soldiers gathered beside a well, with several zoo workers crowding around them, one crying, the others grim and silent.

"How many animals have they already killed?" she asked herself.

Events were unfolding without time to protest or grieve,

and the surviving animals needed help, so she and Jan joined the keepers in feeding, doctoring, and calming animals as best they could.

"At least humans can pack their essentials, keep moving, keep improvising," Antonina thought. "If Germany occupies Poland, what will become of the delicate life-form of the zoo? . . . The zoo animals are in a much worse situation than we are," she lamented, "because they're totally dependent on us. Moving the zoo to a different location is unimaginable; it's too complex an organism." Even if war should erupt and end fast, the aftermath would be costly, she told herself. Where would they find food and money to keep the zoo afloat? Trying not to picture the worst scenario, she and Jan nonetheless bought extra supplies of hay, barley, dried fruits, flour, dried bread, coal, and wood.

On September 7, a Polish officer knocked at the front door and formally ordered all able-bodied men to join the army fighting on a northwestern front—which included forty-two-year-old Jan—and all civilians to vacate the zoo at once. Antonina packed quickly and traveled with Ryś back across the river, this time to stay with her sister-in-law in the west part of the city, in a fourth-floor apartment at No. 3 Kapucyńska Street.

Chapter 5

At night, in the small flat on Kapucyńska street, she learned a new noise: the anvil blows of German artillery. Somewhere else, women her age were slinking into nightclubs and dancing to the music of Glenn Miller, bouncy tunes with names like "String of Pearls" and "Little Brown Jug." Others were dancing to the newly invented jukebox at roadside joints. Couples were hiring babysitters and going to the cinema to see 1939's new releases: Greta Garbo in *Ninotchka*, Jean Renoir's *The Rules of the Game*, Judy Garland in *The Wizard of Oz*. Families were driving through the countryside to view the fall leaves and eat apple cake and corn fritters at harvest festivals. For many Poles, life had become residue, what remains after evaporation drains the juice from the original. During occupation, everyone lost the many seasonings of daily life, trapped in a reality where only the basics mattered and those bled most of one's energy, time, money, and thoughts.

Like other animal mothers, she grew desperate to find a safe hiding place for her young, "but unlike them," she wrote in her diary, "I can't carry Ryś in my jaws to a safe nest." Nor could she remain in her sister-in-law's fourth-floor apartment—"What if the building collapses and we can't escape?"

Maybe it was best, she decided, to resettle downstairs, where a small store sold lampshades—that is, if she could persuade the owners to take her in.

Gathering up Ryś, she climbed down the four flights of dark stairs and knocked on a door which opened to reveal two elderly women, Mrs. Caderska and Mrs. Stokowska.

"Come in, come in." They glanced around the hallway after her and quickly fastened the door.

A strange new continent, half coral reef, half planetarium, veered into view as she entered a cluttered shop redolent with the odors of fabric, glue, paint, sweat, and cooking oatmeal. A bazaar of lampshades hung from the ceiling, nested together in ziggurats or huddled like exotic kites. Wooden shelves held strudel-like bolts of fabric, brass frames, hand tools, screws, rivets, and gleaming trays of finials separated by substance: glass, plastic, wood, metal. In such shops of the era, women sewed new fabric shades by hand, repaired old shades, and sold some made by others.

As Antonina's eyes traveled the room, she would have seen fixtures popular during the 1930s, a time when Baltic decor ran from Victorian to Art Deco and modernist, and included shades such as these: tulip-shaped rose silk decorated with chrysanthemum brocade; green chiffon with lace inset panels of white sateen; geometrically shaped pleated ivory; bright yellow panels in the shape of Napoleon's hat; eight-sided perforated metal with faceted faux jewels inserted around the skirt; dark amber mica crowning a plaster globe embossed with Art Nouveau archers pursuing a stag; a dome of orange-red glass bumpy as gooseflesh, skirted with crystal pendants, below which hung a brass gondola embossed with ivy scrolling. That fashionable red glass, known as *gorge-de-pigeon*, and often used in European wine goblets in Antonina's time, shone sour-cherry red when dark, and when lit, cast a glow the color of freshly peeled blood oranges. It was dyed with pigeon blood, an elixir also used to grade high-quality rubies (with the best stones resembling the freshest blood).

Ryś drew her attention to the far side of the room where, to her surprise, disheveled women and children from the neighborhood sat hedged in by shades.

"Dzień dobry, dzień dobry, dzień dobry," Antonina greeted each woman in turn.

Something about the cozy atmosphere of the lampshade store drew the displaced and bone-chilled to this shop run by grandmotherly ladies willing to share their pantry, coal, and bedding. As Antonina noted,

This lampshade store and workshop was like a magnet to so many people. Thanks to these two tiny lovely old ladies, who were extremely warmhearted, full of love and kindness, we survived this terrible time. They were like the warm light during the summer night, and people from upstairs, homeless people from other locations, from destroyed buildings, even from other streets, were gathering like moths attracted by the warmth around these two ladies.

Antonina marveled as their wrinkled hands passed out food (mainly oatmeal), sweets, a postcard album, and little games. Every night when people chose their spot to sleep, she lay a mattress under a sturdy doorframe and sheltered Ryś with her body, snatching sleep as though falling down a well, as her past grew more idyllic and floated farther away. She had had so many plans for the coming year; now she wondered if she and Ryś would survive the night, if she'd live to see Jan again, if her son would celebrate another birthday. "Every day of our life was full of thoughts of the horrible present, and even our own death," she wrote in her memoirs, adding:

Our allies were not here, not helping us—we Poles were all alone [when] one English attack on the Germans could stop the constant bombing of Warsaw. . . . We were receiving very depressing news about our Polish government—our Marshal Śmigly and members of the government had escaped to Romania and were captured and arrested. We felt betrayed, shocked, we were grieving.

When Britain and France declared war on Germany, Poles rejoiced and radio stations played the French and British national anthems endlessly for days, but mid-September brought no relief from the relentless bombing and heavy artillery. "Living in a city under siege," Antonina wrote disbelievingly in her memoirs, a city full of whistling bombs, jarring explosions, the dry thunder of collapse, and hungry people. First routine comforts like water and gas disappeared, then radio and newspapers. Whoever dared the streets only did so at a run, and people risked their lives to stand in line for a little horsemeat or bread. For three weeks she heard shells zinging over rooftops by day and bombs pounding on walls of darkness at night. Chilling whistles preceded horrible booms, and Antonina found herself listening for each whistle to end, fearing the worst, then letting her breath out when she heard someone else's life exploding. Without trying, she gauged the distance and felt relief that she wasn't the bomb's target, then almost at once came the next whistle, the next blast.

On the rare occasions she ventured out, she entered a film-like war, with yellow smoke, pyramids of rubble, jagged stone cliffs where buildings once stood, wind-chased letters and medicine vials, wounded people, and dead horses with oddly angled legs. But nothing more unreal than this: hovering overhead, what looked at first like snow but didn't move like snowflakes, something delicately rising and falling without landing. Eerier than a blizzard, a bizarre soft cloud of down feathers from the city's pillows and comforters gently swirled above the buildings. Once, long ago, a Polish king repelled invading Turks by attaching large feathered hoops to each soldier's back. As they galloped into battle, the wind coursed through the false wings with a loud tornadic whirring that spooked the enemy's horses, which dug in their hooves and refused to advance. For many Warsawians, this feather storm may have conjured up the slaughter of those knights, the city's guardian angels.

One day, after a live shell plunged into her building and stuck in the fourth-floor ceiling, she waited for an explosion that never came. That night, while bombs sprayed smoke ropes across the sky, she moved Ryś to the basement of a nearby church. Then, "in the strangled silence of the morning," she moved Ryś back again to the lampshade store. "I'm just like our lioness," she told the others, "fearfully moving my cub from one side of the cage to the other."

No news came of Jan, and the worry allowed her little sleep, but she told herself that she would fail him if she didn't save the zoo's remaining animals. Were they even alive, she wondered, and could the teenage boys left in charge really look after them? There seemed no choice: though queasy from fear, she left Ryś with her sister-in-law and forced herself to cross the river amid gunfire and shells. "This is how a hunted animal feels," she thought, caught in the melee, "not like a heroine, just madly driven to get home safely at any cost." She remembered the death of Jaś and the big cats, shot point-blank by Polish soldiers. Visions of their last moments tortured her, and perhaps a fright harder to dispel: What if they turned out to be the lucky ones?

Chapter 6

Nazi bombers attacked Warsaw in 1,150 sorties, devastating the zoo, which happened to lie near antiaircraft guns. On that clear day, the sky broke open and whistling fire hurtled down, cages exploded, moats rained upward, iron bars squealed as they wrenched apart. Wooden buildings collapsed, sucked down by heat. Glass and metal shards mutilated skin, feathers, hooves, and scales indiscriminately as wounded zebras ran, ribboned with blood, terrified howler monkeys and orangutans dashed caterwauling into the trees and bushes, snakes slithered loose, and crocodiles pushed onto their toes and trotted at speed. Bullets ripped open the aviary nets and parrots spiraled upward like Aztec gods and plummeted straight down, other tropicals hid in the shrubs and trees or tried to fly with singed wings. Some animals, hiding in their cages and basins, became engulfed by rolling waves of flame. Two giraffes lay dead on the ground, legs twisted, shockingly horizontal. The clotted air hurt to breathe and stank of burning wood, straw, and flesh. The monkeys and birds, screeching infernally, created an otherworldly chorus backed by a crackling timpani of bullets and bomb blasts. Echoing around the zoo, the tumult surely sounded like ten thousand Furies scratching up from hell to unhinge the world.

Antonina and a handful of keepers ran through the grounds, trying to rescue some animals and release others, while dodging injury themselves. Running from one cage to the next, she also worried about her husband, fighting at the front, "a brave man, a man of conscience; if even innocent animals aren't safe, what hope has he?" And when he returned, what would he find? Then another thought collided: Where was Kasia, the mother elephant, one of their favorites? At last she arrived at Kasia's enclosure, only to discover it leveled and her gone (already killed by a shell, Antonina would later learn), but she could hear two-year-old baby Tuzinka trumpeting in the distance. Many monkeys had died in a pavilion fire or were shot, and others hooted wildly as they scampered through the shrubs and trees.

Miraculously, some animals survived at the zoo and many escaped across the bridge, entering Old Town while the capital burned. People brave enough to stand by their windows, or unlucky enough to be outside, watched a biblical hallucination unfolding as the zoo emptied into Warsaw's streets. Seals waddled along the banks of the Vistula, camels and llamas wandered down alleyways, hooves skidding on cobblestone, ostriches and antelope trotted beside foxes and wolves, anteaters called out *hatchee, hatchee* as they scuttled over bricks. Locals saw blurs of fur and hide bolting past factories and apartment houses, racing to outlying fields of oats, buckwheat, and flax, scrambling into creeks, hiding in stairwells and sheds. Submerged in their wallows, the hippos, otters, and beavers survived. Somehow the bears, bison, Przywalski horses, camels, zebras, lynxes, peacocks and other birds, monkeys, and reptiles survived, too.

Antonina wrote of stopping a young soldier near the villa and asking: "Have you seen a large badger?"

He said: "Some badger banged and scratched on the villa door for a long time, but when we didn't let it in, it disappeared through the bushes."

"Poor Badger," she lamented as she pictured the family pet's frightened appeals at the door. After a moment, "I hope he managed to escape" clouded her mind, the heat and smoke resumed, her legs returned, and she ran to check on the bristle-maned horses from Mongolia. The other horses and donkeys—including her son's pony, Figlarz (Prankster)—lay dead in the streets, but somehow the rare Przywalski horses trembled upright in their pasture.

Antonia finally left the zoo and crossed Praski Park, between rows of linden trees haloed in fire, and headed back to the lampshade store downtown where she and her son sheltered. Blurred and drained, she tried to describe the plumes of smoke, the uprooted trees and grass, the blood-splattered buildings and carcasses. Then, when she felt a little calmer, she made her way to a stone building at No. 1 Miodowa Street and climbed the stairs to a small office crammed with agitated people and cascading piles of documents, one of the Resistance's secret lairs, where she met an old friend, Adam Englert.

"Any news?"

"Apparently, our army is out of ammunition and supplies, and discussing official surrender," he said bleakly.

In her memoirs, she wrote that she heard him speak, but

his words floated away from her; it was as if her brain, already choked by the day's horrors, had issued a *non serviam* and refused to absorb any more.

Sitting down heavily on a couch, she felt glued in place. Until this moment, she hadn't let herself believe that her country might really lose its independence. Again. If occupation wasn't new, neither was ousting the enemy, but it had been twenty-one years since the last war with Germany, most of Antonina's life, and the prospect stunned her. For ten years, the zoo had seemed a principality all its own, protected by the moat of the Vistula, with daily life a jigsaw-puzzle fit for her avid sensibility.

Back at the lampshade store, she told everyone the sorry news she'd heard from Englert, which didn't agree with Polish Mayor Starzyński's upbeat radio broadcasts, in which he denounced the Nazis, offered hope, and rallied everyone to defend the capital at all costs.

"While speaking to you now," he had said on one occasion, "I can see it through the window in its greatness and glory, shrouded in smoke, red in flames: glorious, invincible, fighting Warsaw!"

Puzzled, they wondered whom to believe: the mayor in a public speech or members of the Resistance. Surely the latter. In another broadcast, Starzyński had used the past tense at one point: "I wanted Warsaw to be a great city. I believed that it would be great. My associates and I drew up plans and made sketches of a great Warsaw of the future." In light of Starzyński's tense (was it a slip?), Antonina's news rang truer and everyone's mood fell, as the owners edged among the tables, switching on small lamps.

Several days later, after Warsaw's surrender, Antonina sat at a table with the others, hungry but too depressed to eat the little food in front of her, when she heard a crisp knock at the door. No one visited anymore, no one bought lamps or fixed broken lampshades. Anxiously, the owners opened the door a crack, and to her astonishment there stood Jan, looking exhausted and relieved. Hugs and kisses followed, then he sat down at the table and told them his story.

When Jan and his friends had left Warsaw weeks before, on the evening of September 7, they followed the river and walked toward Brześć on Bug, as part of a phantom army, looking for a unit to join. Not finding one, they finally split up, and on September 25, Jan overnighted in Mienie, at a farm whose owners he knew from summers at the Rejentówka cottage. The following morning, the housekeeper woke him to ask if he'd translate for her with a German officer who had arrived during the night. Any encounter with a Nazi was dangerous, and as Jan dressed he tried to prepare himself for trouble and rehearse possible scenarios. Taking the stairs with the feigned confidence of a legitimate houseguest, he kept his eye on the Wehrmacht officer standing in the living room, discussing provisions with the owners. As the Nazi turned to face him, disbelief washed over Jan, and he wondered if he were seeing something churned up by his jumpy heart. But in the same instant the officer's face flashed surprise and he smiled. There stood Dr. Müller, a fellow member of the International Association of Zoo Directors, who directed the zoo at Królewiec (in eastern Prussia, and known as Königsberg before the war).

Laughing, Müller said: "I know only one Pole well, *you,*

my friend, and I meet you here! How did this happen?" A supply officer, Müller had come to the farm seeking food for his troops. When he told Jan of Warsaw's catastrophe and the zoo's, Jan wanted to return immediately, and Müller offered to help, but warned that Polish men of Jan's age weren't safe on the roads. The best plan, he suggested, was to arrest Jan and drive to Warsaw with him as a prisoner; and despite their past cordiality, Jan worried if Müller could be trusted. But, true to his word, Müller returned when Warsaw surrendered and drove Jan as deep into the city as he dared. Hoping to meet in happier days, they said goodbye, and Jan slid through the ruins of the city, wondering if he'd ever reach Kapucyńska Street, Antonina, and Ryś—if they were even alive. At last he found the four-story building, and when his first knock brought no response, he "nearly toppled from dread."

In the following days, Warsaw's fierce quiet grew unnerving, so Jan and Antonina decided to steal across the bridge to the zoo, this time with no shells or snipers peppering them. Several of the old keepers had also returned and taken up their usual chores as a sort of ghost brigade working in a half-massacred village where the guardhouse and quarters now were charred hills, and the workshops, elephant house, whole habitats and enclosures had also burned or collapsed. Strangest of all, many cage bars had melted into grotesque shapes that looked like the work of avant-garde welders. Jan and Antonina walked to the villa, shocked by a scene that looked even more Surrealist than before. Although the villa had survived, its tall windows were shattered by bomb blasts, and fine particles of glass lay everywhere like sand, mixed with crushed straw from when Polish soldiers had sheltered there during air raids. Everything needed fixing, especially the windows, and because panes of glass were a rare commodity, they decided to use plywood for a while, though it meant sealing themselves off even more.

But first they began a quest for wounded animals, combing the grounds, searching in even unlikely hiding places; a cheer rose whenever someone found an animal, trapped beneath debris, confused and hungry but alive. According to Antonina, many of the army's dead horses lay with swollen bellies, grinning teeth, eyes frozen wide open in fear. All the corpses needed to be buried or butchered (with antelope, deer, and horse meat distributed to the city's hungry), not something Jan and Antonina could face, so they left it to the keepers and at nightfall, exhausted and depressed, the villa uninhabitable, they returned to Kapucyńska Street.

The next day General Rommel spoke on the radio, urging Warsaw's soldiers and citizens to accept surrender with dignity and stay calm while the German army marched into their fallen city. His broadcast ended with: "I rely on the population of Warsaw, which stood bravely in its defense and displayed its profound patriotism, to accept the entry of the German forces quietly, honorably, and calmly."

"Maybe it's good news," Antonina told herself, "maybe it's peace at last and the chance to rebuild."

After a rainy morning, thick cloud banks shifted and a warm October sunlight began streaming through as German soldiers patrolled each neighborhood, filling the streets with the clop of heavy bootheels and gabble in a foreign tongue. Then different sounds filtered into the lampshade store, more sibilant and transparent: crowd voices of Polish men and women. Antonina saw "one large organism flowing slowly" downtown and people trickling out of buildings to join it.

"Where do you suppose they're heading?"

The radio told them where Hitler was preparing to review his troops, and she and Jan felt the same osmotic force tugging them outside. Everywhere Antonina looked lay destruction. In her jottings, she described "buildings guillotined by the war—their roofs gone, sitting in misshapen poses somewhere in nearby backyards. Other buildings looked sad, ripped up by bombs from top to basement." They reminded her of "people embarrassed by their wounds, looking for a way to cover the openings in their abdomens."

Next Antonina and Jan passed rain-soaked buildings missing their plaster, with exposed blood-red bricks steaming in the warm sunshine. Fires still burned, the entrails of homes still smoldered, filling the air with enough smoke to make eyes tear and throats tighten. Hypnotized, the swelling crowd flowed to the center of the city, and in archival films one can see them lining the main streets, down which conquering German soldiers march in a steady torrent of gunmetal-gray uniforms, their steps echoing like ropes walloping hardwood.

Jan turned to Antonina, who looked faint.

"I can't breathe," she said. "I feel like I'm drowning in a gray sea, like they're flooding the whole city, washing away our past and people, dashing everything from the face of the earth."

Jammed inside the crowd, they watched gleaming tanks and guns stream by, and ruddy-faced soldiers, some with stares Jan found so provocative that he had to turn away.

Puppet theater, a popular art form in Poland, wasn't just for children but often grappled with satiric and political subjects, as it had in ancient Rome. Old films show what locals may have found ironic: a loud brass band heralding waves of glossy cavalry and strutting battalions, and Hitler reigning on a platform farther down the avenue, reviewing the troops with one hand held aloft like a puppeteer twitching invisible strings.

Delegates from Poland's main political parties were already meeting in the strong room of a savings bank to refine the Underground, which nearly began with success: explosives planted beneath Hitler's platform were supposed to blow him to crumbs, but at the last minute a German official moved the bomber to another spot and he couldn't light the fuse.

The city quickly spasmed into German hands, banks closed, salaries dried up. Antonina and Jan moved back to the villa, but stripped of money and supplies, they scavenged for food left by the Polish soldiers who had billeted there. The new German colony was ruled by Hitler's personal lawyer, Hans Frank, an early member of the Nazi Party and a leading jurist busy revising German laws according to Nazi philosophy, especially racist laws and those aimed at the Resistance. During his first month in office, Governor-General Frank declared that "any Jews leaving the district to which they have been confined" would be killed, as would "people who deliberately offer a hiding place to such Jews. . . . Instigators and helpers are subject to the same punishment as perpetrators; an attempted act will be punished in the same way as a completed act."

Soon afterward, he issued the "Decree for the Combating

of Violent Acts," which imposed death on anyone disobeying German authority, mounting acts of sabotage or arson, owning a gun or other weapons, attacking a German, violating curfew, owning a radio, trading on the black market, having Underground leaflets in the home—or failing to report scofflaws who did. Breaking laws or failing to report lawbreakers, both acting or observing, were equally punishable offenses. Human nature being what it is, most people didn't wish to get involved, so few people were denounced, and fewer still denounced for not denouncing others . ... in what could quickly have become an absurdist chain of disinclination and inaction. Somewhere between doing and not doing, everyone's conscience finds its own level; most Poles didn't risk their lives for fugitives but didn't denounce them either.

Hitler authorized Frank to "ruthlessly exploit this region as a war zone and booty country, and reduce to a heap of rubble its economic, social, cultural, and political structure." One of Frank's key tasks was to kill all people of influence, such as teachers, priests, landowners, politicians, lawyers, and artists. Then he began rearranging huge masses of the population: over a span of five years, 860,000 Poles would be uprooted and resettled; 75,000 Germans would take over their lands; 1,300,000 Poles would be shipped to Germany as slave labor; and 330,000 would simply be shot.

With courage and ingenuity, the Polish Resistance would sabotage German equipment, derail trains, blow up bridges, print over 1,100 periodicals, make radio broadcasts, teach in covert high schools and colleges (attended by 100,000 students), aid Jews in hiding, supply arms, make bombs, assassinate Gestapo agents, rescue prisoners, stage secret plays, publish books, lead feats of civil resistance, hold its own law courts, and run couriers to and from the London-based government-in-exile. Its military wing, the Home Army, at its height included 380,000 soldiers—among them, Jan Żabiński, who later told interviewers, "from the very beginning, I was connected to the Home Army in the area of the zoo." Confusing as life during occupation must have been, the clandestine Polish state, linked by language rather than territory, would fight nonstop for six years.

A key to the Underground's strength was its no-contact-upward policy and the unflagging use of pseudonyms and cryptonyms. If no one knew his superior, capture wouldn't endanger the core; and if no one knew anyone's real name, saboteurs proved hard to find. Underground headquarters floated around the city, and schools migrated from one church or apartment to another, while a band of couriers and illegal printshops kept everyone informed. The Underground Peasant Movement adopted the slogan of "As little, as late, and as bad as possible," and set about sabotaging deliveries to Germans and diverting supplies to people in the cities, repeatedly claiming delivery of the same grain or livestock, overstating receipts, conveniently losing, destroying, or hiding provisions. Forced laborers in the secret German rocket program at Peenemünde urinated on the electronics to corrode them, crippling the rockets. The Resistance encompassed so many cells that anyone could find a niche, regardless of age, education, or nerve. Jan had a penchant for risk, which he later told a reporter he found exciting, adding in his understated way that its pulse-revving gamble felt rather "like playing chess—either I win or I lose."

Chapter 7

With autumn, cold began seeping under doors and through tiny cracks, and at night hoarse winds dashed across the villa's flat roof, billowed any plywood shutters that had warped, and squalled around the walled terrace. Despite its disheveled buildings and lawns, the zoo bedded down its few remaining animals for winter, but nothing looked the way it had before the war, least of all the seasonal tableaux of zoo life. The tempo of the days used to change dramatically as the zoo entered its own period of hibernation: boulevards, normally crowded with as many as ten thousand people during the summer holiday, would grow almost deserted; a few people would visit the Monkey House, the elephants, the predators' islands, or the seal pool. But the long columns of schoolchildren waiting in line to ride llamas, ponies, camels, or little pedal cars would evaporate. Delicate animals like flamingos and pelicans, venturing outside for a short constitutional each day, would march gingerly in single file over frozen ground. As the days shortened and tree branches grew bare, most animals would stay indoors, while the tone of the zoo paled from raucous to mumbling during what's known in the trade as the *dead season*, a time of animal rest and human repair.

Even in its reduced wartime state, the zoo endured as a complex living machine, where one wobbly screw or stripped gear could trigger a catastrophe, and a zoo director couldn't afford to miss a rusty bolt or a monkey's runny nose, forget to lock or adjust the warmth in a building, overlook a bison's badly matted beard. All this became doubly serious during windstorm, rain, or frost.

Missing now were all the women who used to rake fallen leaves, the men who insulated the roofs and stable walls with straw, the gardeners who wreathed the roses and ornamental shrubs to protect them from frost. Other blue-uniformed helpers should have been cellaring beets, onions, and carrots, and topping off the silos with fodder, so that wintering animals would have plenty of *vitamins* (a word coined in 1912 by Polish biochemist Casimir Funk). The barns should have been brimming with hay, the storerooms and pantries with oats, flour, buckwheat, sunflower seeds, pumpkins, ant eggs, and other essentials. Trucks should have been carting in coal and coke, and the blacksmith fixing broken tools, weaving wire, and oiling padlocks. In the carpentry shop, men should have been repairing the fences, tables, benches, and shelves, and crafting doors and windows for added buildings when the ground softened in spring.

Normally, Antonina and Jan would have been preparing the budget for the coming year, awaiting the arrival of new animals, and reading reports in offices angled to view the river and the steeply roofed houses of Old Town. The press department would have been organizing talks and concerts, lab researchers would have been smearing slides and running tests.

Dead season, though never an easy time of year, usually offered gated asylum in a world private and protected, where they banked on a well-stocked larder, standing orders for foodstuffs, and a belief in self-reliance. The war undermined all three.

"The wounded city is trying to feed their animals," Antonina reassured Jan one morning as she heard a clop-and-clatter, then saw two wagons creaking up to the gate with leftover fruit and vegetable peelings from kitchens, restaurants, and houses. "At least we're not alone."

"No. Warsawians know it's important to save their identity," Jan replied, "all the elements of life that elevate and define them—and, fortunately, that includes the zoo."

Still, Antonina wrote that she felt the ground disappearing beneath her when the occupation government decided to move the capital to Kraków, noting that, as a provincial city, Warsaw no longer needed a zoo. All she could do was await the *liquidation*, a loathsome word suggesting a meltdown of creatures her family knew as individuals, not as a collective mass of fur, wings, and hooves.

Only Antonina, Jan, and Ryś remained at the villa, with not much food at any price, little money, and no jobs. Antonina baked bread every day, and relied on vegetables from the summer garden and preserves made from rooks, crows, mushrooms, and berries. Friends and relatives in outlying hamlets periodically sent food, sometimes even bacon and butter, luxuries seldom seen in the devastated city; and the man who delivered horsemeat to the zoo before the war procured a little meat for them now.

One day in late September, a familiar face appeared at

their front door in German uniform: an old guard from the Berlin Zoo.

"I've been sent directly by Director Lutz Heck with his greetings and a message," he said formally. "He wishes to offer you help, and awaits my call."

Antonina and Jan looked at one another, surprised, not quite sure what to think. They knew Lutz Heck from the annual meetings of the International Association of Zoo Directors, a small clique of altruists, pragmatists, evangelists, and scoundrels. In the early twentieth century, there were two main schools of thought about how to keep exotic animals. One believed in creating natural habitats, the landscape and climate each animal would find in its homeland. The zealous proponents of this view were Professor Ludwig Heck of the Berlin Zoo and his older son, Lutz Heck. The opposing view held that, left to their own devices, exotic animals would adapt to a new environment, regardless of where the zoo was located. The leader of this opposing camp was Professor Lutz's younger son, Heinz, director of the Munich Zoo. Influenced by the Hecks, the Warsaw Zoo was designed to help animals acclimatize, and it also provided inviting habitats. It was the first Polish zoo that didn't cramp animals into small cages; instead, Jan tried to fit each enclosure to the animal, and as much as possible reproduce how it would live in the wild. The zoo also boasted a good natural water source (artesian wells), elaborate drainage systems, and a trained, dedicated staff.

At the annual meetings, ideologies sometimes soured into feuds, but zookeeping families all gloried in their zoos and juggled similar concerns and passions, and thus a freemasonry of shared wisdom and well-being prevailed, despite the language barriers. Other directors didn't speak Polish, Jan didn't speak fluent German, Antonina spoke Polish and some Russian, French, and German. But a sort of Esperanto (a Polish invention) arose that relied heavily on German and English, accompanied by photographs, freehand drawings, animal calls, and pantomime. Annual meetings felt like reunions, and as the youngest zookeeper's wife, Antonina captivated them with her smarts and willowy looks; and they regarded Jan as an energetic and determined director whose zoo was thriving and blessed with rare offspring.

Heck had always been cordial, to Antonina especially. But in his zoo work and now in his politics, he was obsessed with bloodlines, Aryan included, and from what they had heard, he'd become an ardent and powerful Nazi, with Reichsmarshal Hermann Göring and propaganda minister Joseph Goebbels as frequent houseguests and hunting companions.

"We are grateful for Professor Heck's offer," Antonina replied politely. "Please thank him, and tell him that we do not need help, since the zoo is destined to be *liquidated*." She knew full well that, as the highest-ranking zoologist in Hitler's government, Heck might be the very man in charge of the liquidation.

The following day, to their surprise, the guard returned and said that Heck planned a visit soon, and when the guard left they wondered what to do. They didn't trust Heck, but on the other hand, he was sweet on Antonina, and in theory, as a fellow zookeeper, he should be sympathetic to their situation. In an occupied country where survival often depended on having friends in high places, cultivating Heck made sense.

Antonina thought Heck relished the idea of being her patron, a medieval knight like Parsifal, some romantic ideal to win her heart and prove his nobility. As she wondered if his overture signaled help or harm, her mind filled with feline battery: "For all we know he may just be playing with us. Big cats need little mice to toy with."

Jan made a case for Heck's possible goodwill: as a zookeeper himself, Heck loved animals, spent his life protecting them, and undoubtedly sympathized with fellow zookeepers' losses. And so, poised between hope and fear, they passed the night before Lutz Heck's first visit.

After curfew, Poles could no longer stroll under a canopy of stars. They could still watch August's Perseids, followed by the autumn meteor showers—the Draconids, Orionids, and Leonids—from their windows and balconies, but thanks to all the shelling and dust, most days became cloudy with tumultuous sunsets and a drizzle before dawn. Ironically, the far-ranging warfare that created grotesque battlefields and pollution also inspired gorgeous sky effects. Now, fast-falling meteors at night, however kitelike their tails, conjured up images of gunfire and bombs. Once meteors had figured in a category of mind remote from anything technological, as wayfarers from distant realms where stars sparkled like ice-coated barbed wire. Long ago, the Catholic Church had christened the Perseids *the tears of Saint Lawrence* because they occurred near his feast day, but the more scientific image of dirty snowballs pulled by invisible waves from the rim of the solar system, then yanked down to earth, evokes its own saintly magic.

Chapter 8

Lutz Heck took over the Berlin Zoo from his distinguished father in 1931, and almost immediately began remodeling the zoo's ecology and ideology. To coincide with the 1936 Olympics, held in Berlin, he opened a "German Zoo," an exhibit honoring the country's wildlife, complete with "Wolf Rock" at its center, surrounded by enclosures for bears, lynxes, otters, and other native species. This bold patriotic display underscoring the importance of familiar animals, and that one needn't go to the ends of the earth to find exotic species, conveyed a laudable message, and if he'd unveiled his exhibit today, his motives wouldn't be questioned. But given the era, his beliefs, and the ultranationalism of his family, he clearly wanted to please Nazi friends by contributing to the ideal of Germany's master races. A 1936 photograph shows Heck and Göring on a hunting trip to Schorfheide, Heck's large preserve in Prussia; and the following year, Heck joined the Nazi Party.

A big-game hunter, Heck spent the peak moments of his life pursuing danger and adventure, several times a year launching trips to garner animals for his zoo and perhaps bag a pair of longhorn sheep heads for his wall or come face to face with a toweringly mad female grizzly. He relished wild razor-edged hunts, in Africa especially, which he recapped in scenic letters, written by lantern light, astride a camp stool near a well-fed fire, while lions grunted invisibly in the blackness and his companions slept. "The campfire flickering in front of me," he once wrote, "and behind me coming out of the dark infinity the sounds of an invisible and mysterious wild animal." Alone yet faintly haunted by circling predators, he would replay the day's exploits in ink, some to save, some to share with friends in another drape of reality, the Europe that seemed to him planets away. Action photographs often accompanied his letters: lassoing a giraffe, leading a baby rhinoceros, capturing an aardvark, evading a charging elephant.

Heck loved collecting trophies, as memory aids to a wild part of his self that emerged in remote wilderness—live animals to display at his zoo, dead animals to stuff, photographs to share and frame. In the heyday of his travels he seemed to collect life itself, keeping voluminous diaries, snapping hundreds of photographs, and writing popular books (such as *Animals—My Adventure*) which limned his passion for the wilderness, in which he detailed feats of extraordinary bravery, stoicism, and skill. Heck knew his strengths, admired the heroic in himself and others, and could tell a bar-gripping story over drinks at annual meetings. Though he indulged in self-mythology from time to time, his personality fit a profession which has always attracted some people with a yen for exploration, in flight from domesticity, and craving just enough ordeal to feel the threads of mortality fraying. Without his type, maps would still show a flat earth and no one would believe the source of the Nile. Sometimes Heck slayed dragons—or rather their real-life equivalents—but mainly he captured, photographed, and displayed them with gusto. Passionate and single-minded, when he set his sights on an animal, either in the wild or belonging to someone, he lusted after it, tried every lure or ruse he could think of, and persisted until he exhausted the animal or wore down its owner.

For decades the Heck brothers had pursued a fantastic goal, a quest that engaged Heinz but completely infatuated Lutz: the resurrection of three pure-blooded, extinct species—the Neolithic horses known as forest tarpans, aurochsen (the wild cow progenitor of all European cattle breeds), and the European or "forest" bison. On the eve of the war the Hecks had produced some near aurochsen and tarpans of their own, but the Polish strains ran truer to type, the clear inheritors.

Only prehistoric creatures would do, ones untainted by racial mixing, and although Lutz hoped to gain influence and fame in the process, his motives were more personal—he sought the thrill of bringing extinct, nearly magical animals back to life and steering their fate, hunting some for sport. Genetic engineering wouldn't emerge until the 1970s, but he decided to use eugenics, a traditional method of breeding animals which showed specific traits. Heck's reasoning went like this: an animal inherits 50 percent of its genes from each parent, and even an extinct animal's genes remain in the living gene pool, so if he concentrated the genes by breeding together animals that most resembled an extinct one, in time he would arrive at their purebred ancestor. The war gave him the excuse to loot east European zoos and wilds for the best specimens.

As it happens, the animals he chose all thrived in Poland, their historic landscape was Białowieża, and the imprimatur of a respected Polish zoo would legitimize his efforts. When Germany invaded Poland, Heck scouted the farms for mares preserving the most tarpan traits to mate them with several wild strains, including Shetlands, Arabians, and Przywalskis, hoping to breed back to the ideal animal, the fierce, nearly unridable horses painted in ochre on Cro-Magnon caves. Heck assumed it wouldn't take many generations of back-breeding— maybe only six or eight—because as recently as the 1700s tarpans still roamed the forests of northeastern Poland.

During the Ice Ages, when glaciers blanketed northern Europe and a wind-ripped tundra stretched down to the Mediterranean countryside, thick forests and fertile meadows gave refuge to great herds of tarpans that roamed the central European lowlands, browsed the east European steppes, and galloped across Asia and the Americas. In the fifth century B.C., Herodotus said how much he enjoyed watching herds of tarpans grazing in the bogs and marshes of what is now Poland. For ages, purebred tarpans outwitted all the hunters and somehow survived in Europe, but by the eighteenth century not many remained, in part because diners prized tarpan meat—it was sweet, but more appealingly, it was rare—and in part because most tarpans had interbred with farm horses to produce fertile offspring. In 1880, pursued by humans, the last wild tarpan mare fell down a crevasse in Ukraine and died; and the last captive tarpan died seven years later in the Munich Zoo. At that point the species officially became extinct, just one more chapter in the annals of life on Earth.

Humans domesticated wild horses about six thousand years ago, and immediately began refining them: killing the defiant ones for food while breeding the most genial, to produce a horse that submits more easily to saddle and plow. In the process, we revised the horse's nature, compelling it to shed its zesty, ungovernable, evasive *wildness*. The aloof, free-range Przywalski horses retained that fury, and Heck planned to weave their combative spirit into the new tarpan's genetic mix. History credits Colonel Nikolai Przywalski, a Russian explorer of Polish descent, with "discovering" the wild Asiatic horse in 1879, hence its name, though, of course, the horse was well known to the Mongolians, who had already named it *tahki*. Heck factored the *tahki*'s stamina, temper, and looks into his formula, but he craved even older creatures—the horses that dominated the prehistoric world.

What a powerful ideal—that sexy, high-strung horse, pawing the ground in defiance, its hooves all declaratives. Heinz Heck wrote after the war that he and his brother had begun the back-breeding project out of curiosity, but also from "the thought that if man cannot be halted in his mad destruction of himself and other creatures, it is at least a consolation if some of those kinds of animals he has already exterminated can be brought back to life again." But why have tarpans to ride if there were nothing worthy to hunt?

Lutz Heck soon began ministering to a handful of European bison, including those he stole from the Warsaw Zoo, hoping that they might prosper in Białowieża's spirit-house of trees, just as their ancestors had. Heck envisioned forest bison once again galloping along the trails, as sunlight speared through branches of hundred-foot oaks, in a woodland throbbing with wolves, lynx, wild boar, and other game, soon to be joined, he hoped, by herds of ancient horses.

Heck also sought a legendary bull, the aurochs, once the largest land animal in Europe, known for its savagery and vigor. When Ice Age glaciers melted, about twelve thousand years ago, most giant mammals vanished, but in the cold forests of northern Europe, some aurochsen survived, and all modern cattle have descended from those few—not that aurochsen would have been easy to domesticate eight thousand years ago. Because the aurochs went extinct in the 1600s, recent in evolutionary terms, Heck felt sure he could reconstruct it, and in so doing save it, too, from "racial degeneration." He dreamt that, alongside the swastika, the bull might become synonymous with Nazism. Some drawings of the era showed the aurochs and a swastika joined in an emblem of ideological suavity combined with ferocious strength.

Many ancient cultures worshipped the aurochs bull, especially in Egypt, Cyprus, Sardinia, and Crete (whose trans-species ruler supposedly descended from a sacred bull). Zeus often assumed the shape of a bull in Greek myth, the better to ravage alluring mortals and produce offspring with magical gifts; when he abducted Europa, it was in the guise of an aurochs, a great black bull with short beard and giant forward-pointing horns (like those on long-horned cattle, or on the helmets of heroes in the *Nibelungen*). What better totemic animal for the Third Reich? Heck's passion for the project was shared by top Nazi officials, making it clear that Heck's work was not just about the re-creation of extinct species. After Hitler came to power, the biological aims of the Nazi movement spawned many projects to establish racial purity, which justified acts of sterilization, euthanasia, and mass murder. One of the Third Reich's key scientists, Heck's colleague and good friend Eugene Fischer, founded the "Institute of Anthropology, Genetics, and Eugenics," which favored Josef Mengele and other equally sadistic SS doctors who used concentration camp inmates as guinea pigs.

Fascinated by violence and the red-blooded manly spirit— naturally brave, daring, fierce, hardy, sane, lusty, strong-willed—Eugene Fischer believed that mutations in human beings were as destructive as those in domestic animals, and that interbreeding was wilting the human race in the same way that it had already denatured certain "beautiful, good, and heroic" wild animals, losing the potent original in the genetic clutter. The roots of Nazism fed on a lively occultism that spawned the Thule Society, the Germanenorden, the Völkisch movement, Pan-Germanism, and other nationalist cults that believed in a race of Aryan god-men and the urgency of exterminating all inferiors. They exalted superhuman ancestors, whose ancient gnostic rule had brought the Aryans wisdom, power, and prosperity in a prehistoric age until it was supplanted by an alien and hostile culture (namely, Jews, Catholics, and Freemasons); these ancestors were supposed to have encoded their salvation-bringing knowledge in cryptic forms (e.g., runes, myths, traditions), which could be deciphered ultimately only by their spiritual heirs.

This ideal of racial purity really bloomed with Konrad Lorenz, a Nobel scientist highly respected in Nazi circles, who shared Oswald Spengler's belief popularized in *The Decline of the West* (1920) that cultures inevitably decay—but not Spengler's pessimism. Instead he turned to the domestication of animals as an example of how cultures decline, through haphazard breeding of robust and humdrum stock, and championed a biological solution: racial hygienics, a "deliberate, scientifically founded race policy" in which ruin is prevented by the elimination of "degenerate" types. Lorenz used the terms *species*, *race*, and *Volk* interchangeably and warned that "the healthy volkish body often does not 'notice' how it is being pervaded by elements of decay." Describing that decay as the cancer of a physically ugly people and arguing that each animal's goal is the survival of its species, he invoked an ethical commandment he claimed the Bible supported—"Thou shalt love the future of your Volk above all else"—and called for dividing people between those of "full value" and those of "inferior value" (which included whole races and anyone born with mental or physical disabilities), purging the feeble, both in humans and animals.

Heck agreed, aspiring to nothing less than recasting Germany's natural world, cleansing it, polishing it, perfecting it. A true believer from the first stirrings of Nazism, Heck ingratiated himself with the SS, imbibed Fischer's and Lorenz's beliefs about racial purity, and became a favorite with Hitler and, especially, Hermann Göring, his ideal patron. In this sanitary utopia, Heck's job, essentially, was to reinvent nature, and he found Göring a generous patron with deep pockets. In return, Heck wanted to give Göring dominion over Poland's greatest natural treasure, the fantastic lost-in-time preserve on the Polish-Belarussian border, Białowieża. As Heck appreciated, it made the ultimate gift for a man who stamped his coat of arms on most possessions and liked to dress in "pseudo-medieval outfits of long leather jerkins, soft top boots, and voluminous silk shirts, and go marching around his house and estate carrying a spear." Many aristocrats held key positions in the Nazi Party and most of the high command owned hunting lodges or estates, so an important facet of Heck's job was bagging the best hunting preserves and stocking them in novel ways. Dotted with medieval castles, inheritor of Europe's only primeval forest, Poland boasted some of the finest hunting on the continent. Prewar photographs place Göring at his sumptuously appointed hunting lodge northeast of Berlin, on an estate stretching to the Baltic, complete with a 16,000-acre private preserve which he stocked with elk, deer, wild boar, antelope, and other game animals.

More broadly, the Nazis were ardent animal lovers and environmentalists who promoted calisthenics and healthy living, regular trips into the countryside, and far-reaching animal rights policies as they rose to power. Göring took pride in sponsoring wildlife sanctuaries ("green lungs") as both recreation and conservation areas, and carving out great highways flanked by scenic vistas. That appealed to Lutz Heck as it did to many other world-class scientists, such as physicist Werner Heisenberg, biologist Karl von Frisch, and rocket designer Wernher von Braun. Under the Third Reich, animals became noble, mythic, almost angelic—including humans, of course, but not Slavs, Gypsies, Catholics, or Jews. Although Mengele's subjects could be operated on without any painkillers at all, a remarkable example of Nazi zoophilia is that a leading biologist was once punished for not giving worms enough anesthesia during an experiment.

Chapter 9

With blackout in effect and most of the animals gone, dawn no longer announced itself by spilling light into the bedroom and unleashing the zoo's otherworldly chorale. One awoke in darkness and silence, the bedroom windows sealed with plywood and most of the animal calls either missing or muffled. In a quiet that dense, body sounds become audible, one hears blood surging and the bellows of the lungs. In a darkness that deep, fireflies dance across eyes that see into themselves. If Jan were dressing beside the terrace door, Antonina wouldn't have spotted him. If she reached a hand over to the other side of the bed, patted around the pillow, and found it empty, she might have been tempted to loll with memories of zoo life before the war, lost in the dreamy lucidity of her children's books. But on this day, Antonina needed to get busy with her chores, since there were still some animals left to feed, Ryś to dress for school, and the house to prepare for Heck's visit.

Antonina noted that she found Heck "a true German romantic," naïve in his political views and conceited perhaps, but courtly and impressive. She was flattered by his attention, and learned from a mutual friend that she reminded him of his first great love, or so he swore. Their paths rarely crossed, but she and Jan did visit the Berlin Zoo now and then, and Heck had sent them photographs taken on expeditions with cordial letters in which he praised their work.

Antonina slipped into one of several polka-dot dresses she fancied for social occasions (some had a lace or ruffled collar). Photographs almost always show her covered in small lynx-like spots or large pale polka dots against a black or navy blue background that set off her light hair.

From the porch, Jan and Antonina could see Heck's car pass through the main gate—and they no doubt mustered smiles by the time he pulled up.

"Hello, my friends!" Heck said, climbing out. A tall, muscular man with hair combed back and a dark, neatly kept mustache, Heck now wore the uniform of a Nazi officer and the effect was jarring, even if expected, since they were used to seeing him in civilian, zoo, or hunting clothes.

He and Jan shook hands warmly, and he cupped Antonina's hand and kissed it. One can be certain of that, since it was the custom, but not *how* this "true German romantic" might have kissed it. Casually or with a flourish? Lips touching the skin or hovering a breath away? As with a handshake, a hand kiss may reflect subtle feelings—a salute to femininity, a quaking heart, a grudging obedience, a split second of crypto-devotion.

He and Jan would have discussed raising rare animals, particularly those of special interest to Heck, whose life's mission—some would say obsession—dovetailed beautifully with the Nazi desire for purebred horses to ride and purebred animals to hunt.

When it came to rare animals, Jan and Lutz shared a love for those native to Poland, especially the big woolly forest bison (*Bison bison bonasus*), bearded cousin to the North American buffalo (*Bison bison*), and Europe's heaviest land animal. As the recognized expert on these bovines, Jan played a key role in the International Society for the Preservation of the European Bison, founded in Berlin in 1923, with a first agenda of locating all the remaining forest bison in zoos and private collections. It found fifty-four, most beyond breeding age, and in 1932 Heinz Heck traced pedigrees in the first European Bison Stud Book.

Antonina later wrote that as Heck reminisced about their meetings before the war and how much they had in common, once again praising their efforts with the young zoo, she felt hopeful. At last talk turned to the real reason for Heck's visit, which according to Antonina went like this:

"I'm giving you my pledge," he said solemnly. "You can trust me. Although I don't really have any influence over German high command, I'll try nonetheless to persuade them to be lenient with your zoo. Meanwhile, I'll take your most important animals to Germany, but I swear I'll take good care of them. My friends, please think of your animals as a *loan,* and immediately after the war I'll return them to you." He smiled reassuringly at Antonina. "And I will be personally responsible for your favorites, the lynxes, Mrs. Żabińska. I'm positive they'll find a good home in my Schorfheide zoo."

After that, conversation opened to sensitive political topics, including the fate of bomb-ravaged Warsaw.

"At least there's one good thing to celebrate," Heck said, "that the nightmare of September in Warsaw is over and that the Wehrmacht has no further plans to bomb the city."

"What will you do with all your rare animals if war comes?"

"I've been asked that a lot, along with: 'What will you do with the dangerous ones? Suppose your animals escape during an air raid,' and so on. These are terrible thoughts. A vision of Berlin and my zoo after a bombardment by the English is a personal nightmare. I don't want to imagine what might happen to other European zoos if they're bombed. I suppose that's why it grieves me so much to witness your loss, my friends. It's terrible, and I'll do everything I can to help."

"Germany has already turned against Russia. . . ."

"And rightfully so," Heck said, "but overpowering Russia can't happen without England's help, and in the present situation, with England on the other side, our chance of winning is very small."

With so much at stake, Antonina studied Heck carefully. As fleeting emotions stalk it, a face can leak fear or the guilt of a forming lie. The war had a way of curdling her trust in people, but Warsaw's devastation, and the zoo's, clearly rattled Heck. Also, his lack of enthusiasm for Hitler's decisions surprised her, indeed she found "such words, coming from a functionary of the Third Reich, quite shocking." Especially since the Heck she had met before the war rarely shared his political opinions and harped on "German infallibility." Nonetheless, he would soon be shipping her lynxes and other animals to Germany, *to be taken care of*, he'd said, *on loan*, he'd said, and she really had no choice but to comply, stay cordial, and hope for the best.

Chapter 10

The Lutz Heck that emerges from his writings and actions drifted like a weather vane: charming when need be, cold-blooded when need be, tigerish or endearing, depending on his goal. Still, it is surprising that Heck the zoologist chose to ignore the accepted theory of hybrid vigor: that inter-breeding strengthens a bloodline. He must have known that mongrels enjoy better immune systems and have more tricks up their genetic sleeves, while in a closely knit species, however "perfect," any illness that kills one animal threatens to wipe out all the others, which is why zoos keep careful studbooks of endangered animals such as cheetahs and forest bison and try to mate them advantageously. In any case, in the distant past, long before anyone was recognizably Aryan, our ancestors shared the world with other flavors of hominids, and interbreeding among neighbors often took place, producing hardier, nastier offspring who thrived. All present-day humans descend from that robust, talkative mix, specifically from a genetic bottleneck of only about one hundred individuals. A 2006 study of mitochondrial DNA tracks Ashkenazi Jews (about 92 percent of the world's Jews in 1931) back to four women, who migrated from the Near

East to Italy in the second and third centuries. All of humanity can be traced back to the gene pool of one person, some say to a man, some a woman. It's hard to imagine our fate being as iffy as that, but *we* are natural wonders.

Maybe, after decades of observing wild animals, Heck regarded ethnic cleansing as hygienic and inevitable, an engine of reform, replacing one genetic line with an even fitter one, resembling a drama that unfolds throughout the animal kingdom. The usual scenario—using lions as an example—is that an aggressor invades a neighboring pride, kills the lead male and slaughters its young, forcibly mates with the females, thereby establishing his own bloodline, and grabs the previous male's territory. Human beings, gifted at subterfuge and denial yet disquieted by morals, disguise such instincts in terms like *self-defense, necessity, loyalty, group welfare,* etc. Such was the case in 1915, for example, when Turks massacred Armenians during World War I; in the mid-1990s, when Christian Serbs in Bosnia began exterminating the country's Muslims; and in 1994, in Rwanda, when hundreds of thousands of people were slaughtered (and women raped) in warfare between the Hutus and the Tutsis.

The Holocaust was different, far more premeditated, high-tech, and methodical, and, at the same time, more primitive, as biologist Lecomte du Noüy argues in *La dignité humaine* (1944): "Germany's crime is the greatest crime the world has ever known, because it is not on the scale of History: *it is on the scale of evolution.*" That's not to say humans haven't tampered with evolution in the past—we know we've driven many animals to extinction, and we may well have done the same to other lines of humans. Even so, what's instinctive isn't inevitable, we sometimes bridle unruly instincts, we don't always play by nature's rules. No doubt Hitler's twin imperatives of purifying the bloodline while grabbing territory *felt right* along an ancient nerve in people like Heck, to whom it may even have seemed a diabolical necessity.

Heck was also a pragmatist, and Polish lands would soon be re-formed by Germans, zoos included. So when Heck visited the bombed-out Warsaw Zoo, he hid a bleak agenda: his visits were an excuse to loot the finest animals for German zoos and preserves, along with priceless breeding records. Together with his brother Heinz, he hoped to benefit the new German empire and restore the natural environment's lost zest, just as Hitler hoped to reinvigorate the human race.

Repeatedly, Heck swore to the Żabińskis that he had nothing to do with closing their zoo and that his flagging influence with high command wasn't enough to sway generals. Yet Antonina suspected he was lying, that he wielded enormous influence with higher-ups, and might even be personally responsible for their fate. The future of their doomed zoo tortured the Żabińskis, who feared that if it were torn down, plowed up, built on, it would vanish among the casualties of war. Regardless, Jan had to stay at the zoo, whatever that might entail, because it served the Underground, whose foothold in the Praga district in time reached 90 platoons with 6,000 soldiers, the largest pool of saboteurs in the city.

The Home Army, a clandestine branch of the Polish military that took orders from the Polish government-in-exile based in London, fielded a strong hierarchy with a network of scattered cells and many arms dumps, grenade factories, schools, safe houses, messengers, and labs for making weapons, explosives, and radio receivers. As a Home Army lieutenant, Jan sought to disguise the zoo as something the Third Reich might wish to keep intact. The Germans had troops to feed and they loved pork, so he approached Lutz Heck about starting a large pig farm using the ramshackle zoo buildings, knowing that raising pigs in a harsh climate would ensure well-kept buildings and grounds, and even a little income for some of the old staff. According to testimony he gave to the Jewish Historical Institute in Warsaw, by using the ruse of gathering scraps for feeding pigs, he hoped to "bring notes, bacon, and butter and carry messages for friends" in the Ghetto. Antonina wrote:

> We knew that [Heck] was a liar and with great sadness we understood that now there was no hope for saving our zoo. In this situation we decided to talk to Heck about our next plan. Jan wanted to start a large pig farm using our zoo buildings. . . . But we lost our hope about the wild animals in the zoo; Germans were not interested in keeping them alive.

She was right, for although Heck consented to the pig farm, the welfare of the animals not "important" enough for his breeding trials was another matter. First, a noisy caravan of arriving and departing trucks continued for days, hauling the orphan elephant Tuzinka off to Königsberg; shipping the camels and llamas to Hanover; sending the hippopotamuses to Nuremberg; dispatching the Przywalski horses to his brother Heinz, in Munich; and claiming the lynxes, zebras, and bison for the Berlin Zoo. Antonina worried how the upheaval might confuse the animals, which, at journey's end, faced new enclosures, new staff, cajoling or yelling in a new language, new routines, new micro-climates, new mealtimes. Everything would take getting used to, especially new cage-mates and keepers and the sudden loss of herd or family members. All that tumult after the shock of recently being bombed and nearly incinerated. Writing of it, she experienced their suffering twice, as human friend and baffled victim.

After Heck swiped all the animals he wanted for breeding, he decided to host a New Year's Eve shooting party, an old northern European holiday custom based on the pagan belief that noise scares away evil spirits. Traditionally, young men rode from farm to farm, shooting and whooping, banishing the demons, until they were invited indoors for drinks. Sometimes boys circled trees while shooting rifles, ringing bells, and banging on pots and pans, taking part in a timeless ritual designed to rouse nature from her slumber and fill the trees with fruit, the land with a rich harvest.

Warping the tradition, Heck invited his SS friends to a rare treat: a private hunting party right on the zoo grounds, a spree that combined privilege with the pell-mell of exotic animals even a novice or soused gunman could bag. The big-game hunter in Heck coexisted with the naturalist, and paradoxical as it seems, he was a zookeeper who didn't mind killing animals in someone else's zoo if it meant ingratiating himself with powerful friends. Heck and a cadre of fellow hunters arrived on a sunny day, full of drink and hilarity, elated by army victories, laughing as they roamed the grounds, shooting penned and caged animals for sport. Only Göring and his medieval boar spear were missing.

"As a convalescent is hit by a returning fever," Antonina wrote in her diary, "we were hit by the killing of the zoo animals, in cold blood and deliberately on this pretty winter day." Fearing the worst when she saw Heck's friends arrive drunk, jovial, and armed, she decided to keep Ryś indoors.

"Please let me go sledding on the little hill in the llama habitat," he begged. Cooped up all day and cranky, he whined: "I'm bored, and I don't have any playmates."

"How about if we sit in your room and read *Robinson Crusoe*?" she suggested. Reluctantly, he climbed the stairs with her, they curled up on his bed, and she read one of his favorite books by lamplight. But, sensing his mother's gloom, Ryś fidgeted anxiously and couldn't pay attention, even when she reached exciting passages. Suddenly gunshots broke the winter silence, each one followed by its echo, as rifle fire crackled across the grounds, loud enough to hear through shuttered windows.

"Mom, what does it mean?" the frightened boy asked, pulling at her sleeve. "Who is shooting?"

Antonina stared down into the book until its letters began jumping before her eyes, unable to speak or move, hands frozen in place, holding the book's open wings. Dizzying and mutant as the past months were, somehow she had endured, but this moment, "beyond politics or war, of sheer gratuitous slaughter," harrowed her. The savagery didn't serve hunger or necessity, it wasn't a political gambit, the doomed animals weren't being culled because they'd become too abundant in the wild. Not only was the SS ignoring their value as notable creatures with unique personalities, the men didn't even credit animals with basic fear or pain. It was a kind of pornography, in which the brief frisson of killing outweighed the animals' lives. "How many humans will die like this in the coming months?" Antonina asked herself. Seeing and smelling the butchery would have been worse, she wrote, but she found it agonizing to hear shots and imagine the scared animals running, dropping. Her shock, Heck's betrayal, her helplessness dazed her, and she sat paralyzed as her son tugged at her sleeve. If she couldn't protect the animals in her keeping, how could she protect her own son? Or even explain to him what was happening, when the truth would horrify him beyond remedy? Sporadic gunfire continued until late evening, its randomness playing havoc with her nerves, since she couldn't brace herself, only shudder with each shot.

"A very bright, light amaranth sunset was predicting wind for the next day," she wrote later. "Trails, avenues, and frosted yard were covered by thickening layers of snow, which was falling in big chaotic flakes and clusters. In the cold-blue evening light, sunset was playing funeral bells for our just-buried animals. We could see our two hawks and one eagle circling above the garden. When their cage was split open by bullets, they'd flown free, but they didn't want to leave the only home they knew. Gliding down, they landed on our porch and waited for a meal of some horsemeat. Soon even they became trophies, part of the Gestapo officers' New Year's hunting party."

Chapter 11

Life in the zoo stopped cold for weeks, and loss echoed around the cages once filled with familiar snorts and jabbers. Antonina's brain refused to accept the sad new reality, as everywhere a funereal silence hushed the grounds and she tried telling herself that "it wasn't a death sleep but hibernation," the lull of bats and polar bears, after which they would wake refreshed in springtime, stretch their scruffy limbs, and search for food and mates. It was only a rest cure during the icequake and frostbite days of winter when food hid and it was better to sleep in one's burrow, warmed by a storehouse of summer fat. Hibernation time wasn't only for sleep, it was also when bears typically gave birth to cubs they suckled and nuzzled until spring, a time of ripeness. Antonina wondered if humans might use the same metaphor and picture the war days as "a sort of hibernation of the spirit, when ideas, knowledge, science, enthusiasm for work, understanding, and love—all accumulate inside, [where] nobody can take them from us."

Of course, her family's Underground was no sleepy restorative shelter but a policy of hazards, and Antonina found the Underground state of mind a shared "brain-dead reaction" conjured up by the psyche. There was no alternative, really. One needed it to face the stultifying fear and sadness aroused by such daily horrors as people beaten and arrested in the streets, deportations to Germany, torture in Gestapo quads or Pawiak Prison, mass executions. For Antonina, at least, that flight, stoicism, or dissociation—whatever one labels it—never quite dispelled the undertow of "fear, rebellion, and extreme sadness."

As Germans systematically reclaimed Polish towns and streets, even speaking Polish in public became forbidden; in Gdańsk it was punishable by death. The Nazi goal of more "living space" (*Lebensraum*) applied pointedly to Poland, where Hitler had ordered his troops to "kill without pity or mercy all men, women, and children of Polish descent or language. Only in this way can we obtain the *Lebensraum* we need." Those children thought to show the strongest Nordic features (and thus genes) were destined for Germany to be renamed and raised by Germans. Like the Hecks, Nazi biologists believed in appearances, that anyone who strongly resembled a target species could be bred back to a pure ancestor.

The racial logic went like this: A biologically superior Aryan race had spread across the world, and though various empires collapsed, traces of Aryans remained among the nobility, whose features could be identified and harvested from descendants in Iceland, Tibet, Amazonia, and other regions. Working on this theory, in January 1939, Reichsführer Himmler launched a German Tibet Expedition to locate the roots of the Aryan race, led by twenty-six-year-old naturalist, hunter, and explorer Ernst Schäfer.

"Himmler shared at least one passion with Ernst Schäfer,"

Christopher Hale writes in *Himmler's Crusade*: he "was fascinated by the East and its religions," going so far as to carry
a notebook "in which he had collected homilies from the
Hindu *Bhagavadgita* ('Song of the Lord'). To the unimpressive
little man [Himmler] who sat inside the poisonous spider's
web of the SS, Ernst Schäfer was an emissary from another
mysterious and thrilling world." Himmler also brewed a deep
hatred for Christianity, and since most of Poland was devoutly
Catholic, all Poles drew punishment.

Antonina wrote that her world felt gutted, collapsing in
slow motion, and that for a *Blitzkrieg,* a lightning-fast war,
"it had many long-drawn-out phases." Food stamps entered
their lives and costly black-market food, though luckily
Antonina could still bake bread from the grain she had bought
from her sister-in-law in the fall.

At winter's end, she and Jan started receiving the first
shipments of sows, and by March of 1940 the pig farm began,
mainly fed on scraps donated from restaurants and hospitals,
as well as garbage Jan collected in the Ghetto. Grossly over-
qualified, the old keepers looked after the pigs and the animals
thrived, producing several hundred piglets during the summer,
which provided the household with meat and served Jan's
main objective of using the zoo as an Underground depot.

One spring day, Jan brought home a newborn piglet whose
mother was just butchered, thinking that Ryś might like it
as a pet, and Antonina found him a bristly scramble of energy,
hard to bottle-feed, especially when he started gaining weight.
They named him Moryś, and at two and a half weeks, Moryś
looked like "a piglet from *Winnie-the-Pooh* . . . very clean, pink
and smooth, with a marzipan beauty," she wrote. (In Poland,

children usually received little pink marzipan pigs for Easter.)

Moryś lived in the so-called attic of the villa, really a long narrow closet that shared a terrace with the upstairs bedrooms, and each morning Antonina found him waiting outside Ryś's bedroom door. When she opened it, Moryś "ran into his room, oinking, and started jostling Ryś's hand or foot until Ryś woke, stretched out a hand, and scratched Moryś's back. Then the pig arched, catlike, until he looked like the letter *C*, and grunted with great contentment," uttering a quiet noise between a snort and a creaking door.

On rare occasions Moryś risked going downstairs into a stew of smells and voices, a maze of strange human and furniture legs. The clinking of a dinner table being set usually lured him to the top of the staircase, where he parked himself and "blinked his buttery blue eyes with long white eyelashes, looking and listening," Antonina wrote. If someone called him, he edged down the polished wooden stairs, carefully, hooves slipping now and then, skittered into the dining room, and circled the table, hoping for a handout, though scraps were few.

After dinner each evening, Moryś and Ryś repaired to the garden to gather grass and weeds to feed the rabbits living in the old Pheasant House, which gave Moryś a chance to hunt for tubers and greens. That scene incandesced in Antonina's memory, the icon of her little boy and his pig playing in the lavender twilight: "Ryś and Moryś on a field of green, which captivated everyone. Watching them, we could forget the war's tragic events for long moments." Her son had lost so much childhood, so many pets, including a dog, a hyena pup, a pony, a chimpanzee, and a badger, that Antonina cherished his daily flights with Moryś into the vegetable garden's vest-pocket Eden.

One puzzle of daily life at the villa was this: How do you retain a spirit of affection and humor in a crazed, homicidal, unpredictable society? Killers passed them daily on zoo grounds, death shadowed homely and Underground activities alike, and also stalked people at random in the streets. The idea of safety had shrunk to particles—one snug moment, then the next. Meanwhile, the brain piped fugues of worry and staged mind-theaters full of tragedies and triumphs, because unfortunately, the fear of death does wonders to focus the mind, inspire creativity, and heighten the senses. Trusting one's hunches only seems a gamble if one has time for *seem*; otherwise the brain goes on autopilot and trades the elite craft of analysis for the best rapid insights that float up from its danger files and ancient bag of tricks.

Chapter 12

"How can this barbarity be happening in the twentieth century?!!!!!!" Antonina asked herself, an outcry of disbelief with no fewer than six exclamation points. "Not long ago the world looked on the dark ages with contempt for its brutality, yet here it is again, in full force, a lawless sadism unpolished by all the charms of religion and civilization."

Sitting at the kitchen table, she prepared small packets of food for friends in the Ghetto, thankful that no one poked through Jan's clothing or pails as he went about his regular rounds to collect kitchen scraps for the Weimar's pig farm. No doubt he enjoyed the irony of carrying food *from* the pig farm *into* the Ghetto, and if it felt a little off-color giving Jews pork, a taboo food, dietary laws had long since been waived, and everyone was grateful for protein, a scarce gift on either side of the wall.

In the beginning, neither Jews nor Poles absorbed the full tirade of racist laws or believed the grisly rumors about Jew roundups and killings. "As long as we didn't witness such events themselves, feel it with our own skin," Antonina later recalled, "we could dismiss them as otherworldly and unheard-of, only cruel gossip, or maybe a sick joke. Even when a

Department of Racial Purity opened a detailed census of the city's Jewish population, it still seemed possible to attribute such madness to that famous German talent for being systematic and well organized," the wheel-spinning of bureaucrats. However, Germans, Poles, and Jews stood in three separate lines to receive bread, and rationing was calculated down to the last calorie per day, with Germans receiving 2,613 calories, Poles 669 calories, and Jews only 184 calories. In case anyone missed the point, German Governor Frank declared: "I ask nothing of the Jews except that they disappear."

Verboten! became a familiar new command, yelled by soldiers, or inked large with the wagged finger of an exclamation point on posters and in anti-Semitic newspapers like *Der Stürmer*. Ignoring those three syllables was punishable by death. Barked out, the word moved from fricative *f* to plosive *b*, from thin-lipped disgust to blown venom.

As warnings and humiliations increased day by day, Jews were forbidden restaurants, parks, public toilets, and even city benches. Branded with a blue star of David on a white armband, they were barred from railways and trams, and publicly stigmatized, brutalized, denigrated, raped, and murdered. Edicts forbade Jewish musicians from playing or singing music by non-Jewish composers, Jewish lawyers were disbarred, Jewish civil servants fired without warning or pension, Jewish teachers and travel agents dismissed. Jewish-Aryan marriages or sexual relations were illegal, Jews were forbidden to create art or attend cultural events, Jewish doctors were ordered to abandon their practices (except for a few in the Ghetto). Street names that sounded Jewish were rechristened, and Jews with Aryan-sounding first names had to replace them with "Israel" or "Sarah." Marriage licenses issued to Poles required a "Fitness to marry" certificate. Jews couldn't hire Aryans as servants. Cows couldn't be inseminated by Jewish-owned bulls, and Jews weren't allowed to raise passenger pigeons. A host of children's books, like *The Poison Mushroom*, promoted Nazi ideology with anti-Semitic caricatures.

For sport, soldiers hoisted orthodox Jews onto barrels and scissored off their religious beards, or taunted old men and women, sometimes ordering them to dance or be shot. Archival footage shows strangers waltzing together in the street, holding each other awkwardly, faces sour with fear, as Nazi soldiers clapped and laughed. Any Jew passing a German without bowing and doffing his cap merited a savage beating. The Nazis seized all cash and savings, and stole furniture, jewelry, books, pianos, toys, clothing, medical supplies, radios, or anything else of value. Over 100,000 people, yanked from their homes, endured chronic days of physical labor without pay, and Jewish women, as further humiliation, were forced to use their underwear as cleaning rags on floors and in toilets.

Then, on October 12, 1940, the Nazis ordered all of Warsaw's Jews from their homes and herded them into a district on the north side of town, which lay conveniently between the main railway station, Saxon Garden, and the Gdańsk railroad terminal. Typically, German soldiers would surround a block and give people half an hour to vacate their apartments, leaving everything behind but a few personal effects. Adding the Jews relocated from the countryside, that edict confined 400,000 people to only 5 percent of the city, about fifteen to twenty square blocks, an area about the size of Central Park, where the sheer racket alone, a "constant tense clamor" as one resident described it, frayed sanity. That vortex of 27,000 apartments, where an average of fifteen people shared two and a half small rooms, served the Nazi goal of grinding down morale, enfeebling, humiliating, and softening up resistance.

Jewish Ghettos had flourished in Europe throughout history, and however remote or disdained, they tended to be vital and porous, allowing travelers, merchants, and culture to flow in both directions. The Warsaw Ghetto differed dramatically, as Michael Mazor, a Ghetto survivor, recalls: "In Warsaw the Ghetto was no longer anything but an organized form of death—a 'little death chest' (*Todeskätschen*), as it was called by one of the German sentries posted at its gates . . . a city which the Germans regarded as a cemetery." Only the crafty and vigilant survived, and no one ventured from home without first checking the danger forecast. Pedestrians updated one another as they passed, and "mere mention of a threat, the slightest gesture, could send a crowd of several thousand back inside, leaving the street empty and bare."

But life's weedy tumult still flowered in the Ghetto, however and wherever it could. Norman Davies gives this snapshot of the early Ghetto's vibrant features: "For two or three years, it was thronged with passers-by, with rickshaws and with its own trams mounted with a blue star of David. It had cafes and restaurants, at number 40 a 'Soup Kitchen for Writers,' and places of amusement. The Fotoplastikon at 27 Leszno Street offered a popular eye on the outside world by showing a series of still pictures of exotic places like Egypt, China, or California.

A clown with a red nose stood on the pavement, cajoling people to buy a ticket for 6 groszy. At 2 Leszno Street, the Arts Coffee House laid on a daily cabaret and a stream of concerts featuring singers such as Vera G. or Marysdha A., the 'Nightingale of the Ghetto,' and musicians such as Ladislas S. and Arthur G. At 35 Leszno Street, the 'Femina' music hall mounted more ambitious productions from a wide Polish repertoire including the 'Princess of the Czardas' revue, and the aptly named comedy 'Love Seeks an Apartment.' It was all a desperate form of escapism. As someone remarked, 'Humor is the Ghetto's only form of defence.' " Many of the Ghetto's best-known streets translated as visions of paradise, plenty, and adventure: Garden Street, Peacock Street, Cool Street, Wild Street, New Linden Street, Dragon Street, Salt Street, Goose Street, Brave Street, Warm Street, Cordials Street, Pleasant Street.

At first, while the Ghetto remained porous, the Żabińskis' Jewish friends believed it a temporary lepers' colony, or that Hitler's regime would quickly collapse and justice prevail, or that they could weather out the maelstrom, or that the "final solution" meant ejecting Jews from Germany and Poland—anything but annihilation.

Choosing an unknown future over a violent present, most Jews moved as ordered, though some, contrary and beyond herding, opted for a chancy life hiding on the Aryan side of the city. According to Antonina, a bleak topic of conversation among her friends of mixed descent, or married couples one of whom was Jewish, were the racist Nuremberg laws of September 15, 1935, stipulating how much Jewish blood you could have without being tainted. The famous explorer of the Silk Road and Nazi apologist Sven Hedin, who stood beside

Hitler on the podium at the 1936 Olympics, was exempt, though his great-grandfather had been a rabbi, something Hitler's inner circle would surely have known.

Although few people foresaw the racist laws as a matter of life and death, some quickly converted to Christianity and others bought false documents. Afraid the Germans might discover Wanda's part-Jewish heritage, their friends Adam and Wanda Englert arranged a fake divorce followed by a nonevent known as "Wanda's Disappearance." But before Wanda vanished, she decided to throw a farewell party for family and close friends at the old armory downtown, and she chose summer solstice for the event.

On this holy eve, the armory was undoubtedly decorated with sprigs of mugwort, a tall plant in the wormwood family with purplish stems, gray-green leaves, and small yellow flowers. The ancient herb was used to break spells and repel male and female witches, especially on June 23, Midsummer's Eve, a day associated with Saint John (according to legend, when Saint John was beheaded, his head tumbled into a patch of mugwort plants). Superstitious Polish farmers hung branches of the herb under barn eaves to keep witches from milking the cows dry during the night, Warsawian girls wore mugwort garlands in their hair, and housewives tied sprigs of mugwort to doorways and windowsills to dash evil. During occupation by perceived devils, a party held on Midsummer's Eve couldn't have been a coincidence.

On June 22, Jan and Antonina set out for the party, planning to cross Kierbedź Bridge, a pleasant stroll or trolley ride in good weather. In old photographs, the bridge's enclosed metal trusses look like a long row of staples, and its basket-weave stencils the road with small squares of sunlight. Such bridges flute tunelessly when wind pipes through at changing speeds, and vibrate with felt music, a bone-buzzing bass also made by elephants, who speak and hear in subsonic, which zookeepers can feel if they stand where elephants talk.

Jan and Antonina usually took a shortcut through Praski Park, whose urban oasis once spread to seventy-four acres over old Napoleonic fortifications. In 1927, the new zoo absorbed about half of the park, leaving in place as many old trees as possible, so that people arriving by trolley first passed beneath the arbors to find the zoo unfolding with the same species of honey locusts, maple-leaved sycamores, maiden-hairs, and sweet chestnut trees as prologue and story. But on this afternoon, discovering they were out of cigarettes, Jan and Antonina chose a longer route, along Łukasiński Street, which skirted the park, and popped into a little shop full of the sweet smell of strong Polish tobacco. Just as they were leaving and lighting up, a great booming shock wave hurled them against a fence and rocks rained through a cloud of sandy soil. At once the air curdled and turned black, and a second later they heard an airplane engine and saw a thin pink line streaking across the sky. Their lips moved without sound as they staggered to their feet, deafened and confused by the blast. Then, when wolf-howl sirens blew an all-clear, they decided the plane wasn't part of a wave but a lone bomber trying to destroy Kierbedź Bridge, which remained intact, as did Praski Park. But a spume of black smoke gusted and rose and gusted again from a blasted trolley.

"If we'd taken the shortcut, we might have been on it," Jan said angrily.

A second fright gripped Antonina as she noticed the time. "But that *is* the trolley Ryś sometimes takes home from school!"

Sprinting down the street, they ran to the sparking, twitching trolley, tossed from its tracks and lying in front of the Catholic church like a steaming mammoth, its metal mangled and wire umbilicals lax, with fifty or so limp people scattered inside and out. "With tears rolling from my eyes, I looked into the faces of the dead, looking for Ryśio's face," Antonina recalled. Searching through smoke and hot debris for their son, and not finding him, they ran to the school, but the children had already left. Next they ran back past the trolley and swelling crowd, through Praski Park, rushing between the cages to the villa, racing up the back steps, bursting into the kitchen, and hunting the whole house, shouting Ryś's name.

"He's not here," Jan said at last, sagging into a chair. After a while, they finally heard him on the back steps.

"Sit down," Jan said sharply but quietly as he steered Ryś to a chair. "Where did you go, you bad boy? Did you forget that returning home from school at once is your chief responsibility?"

Ryś explained that school had just let out when a bomb hit, and then a worried stranger had herded the children inside his house until the all-clear siren blew.

Needless to say, Antonina and Jan missed Wanda's party, but not her company, because soon afterward, as planned, she "disappeared" to the zoo, in the guise of Ryś's non-Jewish tutor.

Chapter 13

Jan and Antonina found Nazi racism inexplicable and devilish, a disgust to the soul, and although they were already assisting friends inside the Ghetto, they pledged, despite the hazards, to help more Jews, who had figured importantly in Jan's childhood memories and loyalties.

"I had a moral indebtedness to the Jews," Jan once told a reporter. "My father was a staunch atheist, and because of that, in 1905, he enrolled me in the Kretshmort School, which at that time was the only school in Warsaw where the study of Christian religion wasn't required, even though my mother was very opposed to it because she was a devout Catholic. Eighty percent of the students were Jews, and there I developed friendships with people who went on to distinguish themselves in science and art. . . . After graduating high school, I began teaching in the Roziker School," also predominantly Jewish. As a result, he made intimate friendships among the Jewish intelligentsia, and many school chums lived behind Ghetto walls. Although Jan didn't say much publicly about his father, he told a journalist that he'd chosen zoology "to spite my father, who didn't like or appreciate animals, and didn't allow them in the house—other than moths and flies, who entered without his permission!"

They had more in common when it came to the loyalty shown Jewish friends:

My father and I both grew up in a Jewish neighborhood. He was a lawyer, and even though he married into a very wealthy family—the daughter of a landowner—he rose to bourgeois status on his own. It was just by chance that we happened to grow up in this poor Jewish neighborhood in Warsaw. From childhood my father used to play with Jewish children in the streets, treating Jews as equals. And I was influenced by him.

The zoo was by no means ideal for hiding refugees. The villa stood close to Ratuszowa Street, right out in the open like a lighthouse, surrounded only by cages and habitats. A clutch of houses for employees and administrative buildings lay mid-zoo, three-tenths of a mile away; acres of open land encircled the villa, most of it a park with small garden plots; railroad tracks ran south, along the Vistula River, just beyond the zoo fence; and the north side held a military zone of small wooden buildings heavily guarded by German soldiers. After Warsaw's surrender, on the lions' island right in the middle of the zoo, Germans had built a storehouse for weapons confiscated from the Polish army. Other German soldiers often visited the zoo as well, for a dose of greenery and quiet, and no one could predict how many might appear, or when, since they didn't seem to favor one time of day over another. But they arrived in an off-duty frame of mind, not on patrol, and, in any case, Praski Park's less-bombed setting offered more appealing walks.

Amazingly, Antonina never twigged one of Jan's secrets: that with his help the Home Army kept an ammunitions dump at the zoo, buried near the moat in the elephant enclosure. (A small paneled room was found there after the war.) He knew the danger, even foolhardiness, of burying guns right in the center of the zoo, steps away from a German military warehouse, but how could he tell her? He worried that she'd be terrified and insist the family's safety came first. Luckily, as Jan thought, it never occurred to the Germans that a Pole would be that gutsy, because they regarded Slavs as a fainthearted and stupid race fit only for physical labor.

"Knowing the German mentality," he reasoned, "they would never expect any kind of Underground activity in a setting so exposed to public view."

Jan always shied away from praise and underplayed his bravery, saying such things as: "I don't understand all the fuss. If any creature is in danger, you save it, human or animal." From interviews, his own writings, and Antonina's accounts, he comes across as naturally private yet sociable, highly disciplined, strict with himself and his family, the sort of man we sometimes call "a cool customer," gifted with the ability to hide his deeds and feelings, someone with enormous *hart ducha* (strength of will or spirit). In the Polish Underground, where acrobatic feats of daring unfolded daily, Jan bore the code name "Francis," after Francis of Assisi, patron saint of animals, and was known for his audacity, sangfroid, and risk-taking. His choosing to hide weapons and Jews in plain sight, in the heart of a Nazi encampment, proved to be good psychology, but I think it was also a kind of one-upmanship he savored, a derisive private joke. Still, discovery would have meant pitiless, on-the-spot death for him and his family, and who knows how many others. Creating a halfway house, "a stopping place for those who escaped the Ghetto, until their destinies were decided and they moved to new hideouts," Jan discovered that being an atheist didn't shield him from a robust sense of fate and his own personal destiny.

Chapter 14

In the summer of 1940, a phone call, a note, or a whisper might alert the Żabińskis to expect secret "Guests" placed by the Underground. Jews in hiding and in transit, nomads, not settlers, they stopped briefly to rest and refuel en route to unnamed destinations. German-speaking Jews who looked Aryan received false identity papers and sailed smoothly through, and those who couldn't pass spent years at the zoo, some in the villa and as many as fifty at a time in empty cages. Many Guests, like Wanda Englert, were longtime friends or acquaintances, and Antonina regarded them as one amphibious family. Hiding them posed problems, but who better than zookeepers to devise fitting camouflage?

In the wild, animals inherit clever tricks of blending into their surroundings; for instance, penguins are black on top and white on the bottom so that the patroling skuas will assume they're a twist of ocean and leopard seals dismiss them as clouds. The best camouflage for people is more people, so the Żabińskis invited a stream of legal visitors—uncles, aunts, cousins, and friends for varying stays—and established a regular unpredictability, a routine of changing faces, physiques, and accents, with Jan's mother a frequent guest.

"Everybody loved Jan's mother," Antonina noted in her memoirs. "She had a kind, graceful nature, and she was very smart, a fast thinker with an excellent memory, very polite and sensitive. She had a big full-bodied laugh and a great sense of humor." But Antonina did worry about her, because "she's like a delicate greenhouse flower, and our duty was to protect her from any fear or pain that might damage her spirit or trigger a depression."

Jan left those intangibles to Antonina, who always handled the "difficult animals" and for whom the chance to amuse, impress, and, ultimately, rescue a parent surely appealed in visceral ways. Jan preferred the role of general, spy, and tactician, especially if it meant bamboozling or humiliating the enemy.

Unlike other occupied countries, where hiding Jews could land you in prison, in Poland harboring a Jew was punishable by immediate death to the rescuer and also to the rescuer's family and neighbors, in a death-frenzy deemed "collective responsibility." Nonetheless, many hospital workers disguised adult Jews as nurses, drugged small children to quiet them before smuggling them out in knapsacks, and planted people in funeral carts under a heap of corpses. Many Christian Poles hid Jewish friends for the whole length of the war, even though it meant reduced rations and relentless vigilance and ingenuity. Any extra food entering the house, unfamiliar silhouettes, or whispers seeping from a cellar or closet might inspire a visiting neighbor to notify the police or tip off the city's underbelly of blackmailers. The wayfarers often spent years in the dark, barely able to move, and when they finally emerged, unfolding their limbs, their weak muscles failed and they needed to be carried like a ventriloquist's dummies.

The zoo wasn't always a first stop for Guests, especially ones escaping the Ghetto, who might spend a night or two downtown with Ewa Brzuska, a short, ruddy, squarish woman in her sixties whom people called "Babcia" (Granny). She owned a tiny grocery (sixteen feet by three feet) on Sędziowske Street, which extended out onto the sidewalk where Ewa arranged barrels of sauerkraut and pickles beside baskets of tomatoes and greens. Neighbors crowded to shop and socialize, despite the German military's car repair depot right across the road. Every day, a group of Jewish men would be escorted from the Ghetto to work on the cars, and Granny would secretly post their letters or keep watch while they spoke with family members. Tall sacks of potatoes stood around for young smugglers from the Ghetto to hide behind. In 1942, her back rooms became a branch office of an Underground cell, and she stored ID cards, spare birth certificates, money, and bread coupons under barrels of pickled cucumbers and sauerkraut, stashed subversive publications in the stockroom, and often hid escaping Jews for a night, some surely bound for the zoo.

Antonina rarely knew when to expect Guests, or where they came from; Jan handled the plotting and liaised with the Underground, and as a result, no one hiding in the villa guessed the full measure of his Underground activities. They didn't know, for example, what was hidden inside the Nestlé or Ovaltine boxes which would appear from time to time on the shelf above a radiator in the kitchen.

Antonina reports Jan saying casually one day: "I put some

small springs for my research instruments into this box. Please don't touch or move it. I may need it at any time."

No one raised an eyebrow. Jan had always been a collector of small metal findings—screws, washers, and gizmos—though he usually stored them in his workshop. Those who knew him found his hobby quaint, a hardware junkie's pastime. Not even Antonina realized that he was collecting fuses for making bombs.

When a young researcher from the Zoological Institute arrived with a big barrel of fertilizer, Jan stashed it in the animal hospital next to the villa, and every now and then he'd mention in passing that so-and-so might come to fetch some fertilizer for his garden. Antonina only learned after the war that the barrel actually contained C13F, a water-soluble explosive, and that Jan was the leader of an Underground cell that specialized in sabotaging German trains by jamming explosives into wheel bearings, so that when the train started to move, the powder would ignite. (During one month in 1943, they derailed seventeen trains and damaged one hundred locomotives.) She didn't know during the war that he also infected some pigs with worms, butchered them, then shaped the poisoned meat into balls which, with the help of an eighteen-year-old working in a German army canteen, he slipped into the soldiers' sandwiches.

He also helped to build bunkers, vital underground dens. In wartime Poland, the word *bunker* didn't conjure up the simple trench it might today, but a damp underground shelter with camouflaged shafts and air vents, usually located at the edge of a garden or public park. Emanuel Ringelblum's bunker at 81 Grójecka Street, lying under a market gardener's greenhouse, ran ninety-two feet square and housed thirty-eight people on fourteen crowded beds. One of his bunker-mates, Orna Jagur, who, unlike Ringelblum, left the bunker before it was discovered in 1944, recalls the moment she first inhaled bunker life:

> A wave of hot stuffy air struck me. From below there poured out a stench made of mildew mixed with sweat, stale clothing, and uneaten food. . . .
>
> Some of the inhabitants of the shelter were lying on the bunks, sunken in darkness, the rest were sitting at the tables. Because of the heat, the men were half-naked, wearing only pajama bottoms. Their faces were pale, tired. They had fear and unease in their eyes, their voices were nervous and strained.

That was considered a well-built bunker tended by a caring family who provided decent food, an unusually good hideout.

By comparison, life at the zoo seemed roomy and bucolic, if zany, and people in the Underground referred to it by cryptonym, as "The House Under a Crazy Star," more an oversized curiosity cabinet than a villa, where the lucky escaped notice among a hodgepodge of eccentric people and animals. Urban visitors relished the futuristic villa with the large park embracing it, offering forty or so acres of green vistas where they could forget the war and pretend to be vacationing in the country. Since paradise only exists as a comparison, Guests in flight from the Ghetto found villa life a small Eden, complete with garden, animals, and motherly bread-maker (the etymological origin of the word *paradise*).

After dark, by official order, the Żabińskis hung black paper over the windows, but by day the two-story, supposedly one-family, villa pulsed like a beehive behind glass. With all the legal residents on board—housekeeper, nanny, teacher, in-laws, friends, and pets—mingling silhouettes and weird noises seemed normal. Startlingly visible, the villa shone like a display box, with a few low shrubs growing around it, some mature trees, and its signature tall windows. Jan staged things that way on purpose, with full exposure and lots of human traffic, abiding by the axiom *more public, less suspicious.*

Why so much glass? The villa showcased the International Style of architecture, a mode that ignored the history, culture, geology, or climate surrounding a house. Instead, with a bow to the machine age and Futurism, it strove for radical simplicity, without ornamental features, in sleek buildings constructed from glass, steel, and concrete. Architectural leaders—Walter Gropius, Ludwig Mies van der Rohe, Marcel Breuer, Le Corbusier, and Philip Johnson—hoped to reflect honesty, directness, and integrity by creating open-faced buildings with nothing to hide. The movement's slogans said it all: "ornament is crime," "form follows function," "machines for living." At odds with Nazi aesthetics (which worshipped classical architecture), building and living in a modernist villa was itself an affront to National Socialism, and Jan and Antonina made the most of all the style implied: transparency, honesty, simplicity.

In that flux, where people appeared and disappeared, unnamed and unexpected, it proved hard to spot the Guests, even harder to tell which people were not there, and when. However this crypto-innocence meant living on the edge and silently glossing every noise, tracking every shadow. Did a sound fit into the ever-changing concordance of villa life? Inevitably, a vital paranoia reigned in the house as the only sane response to perpetual danger, while its inhabitants mastered the martial arts of stealth: tiptoe, freeze, camouflage, distract, pantomime. Some villa Guests hid while others hovered, emerging only after dark to roam the house at liberty.

So many people also meant added chores for Antonina, who had a large family to oversee; livestock, poultry, and rabbits to manage; a busy garden with tomatoes and pole beans to stake; bread to bake daily; preserves, pickled vegetables, and compotes to jar.

Poles were growing used to occupation's unexpected frights, finding the pulse calm one moment and sprinting the next as war reset their metabolism, especially the resting level of attention. Each morning, they awoke in darkness, not knowing the day's fate, maybe sorrowful, maybe ending in arrest. Would she be one of those people, Antonina wondered, who vanished because they happened to be on a tram or in a church when Germans chose it at random, sealed off the exits, and killed everyone inside as revenge for some real or imaginary insult?

Household chores, however humdrum and repetitious, lulled with motions familiar, harmless, and automatic. Constant vigilance had become exhausting, the senses never quite relaxed, the brain's watchmen kept patrolling the wharves of possibility, peering into shadows, listening for danger, until the mind became its own penitent and prisoner. In a country under a death sentence, with seasonal cues like morning light or drifting constellations hidden behind shutters, time changed shape, lost some of its elasticity, and Antonina wrote that her days grew even more ephemeral and "brittle, like soap bubbles breaking."

Soon Finland and Romania sided with Germany, and Yugoslavia and Greece surrendered. Germany's attack on its former ally, the Soviet Union, triggered rounds of rumors and forecasts, and Antonina found the Battle of Leningrad especially depressing, since she'd hoped the war might be winding down, not flaming anew. At times, she heard that Berlin had been bombarded, that a Carpathian brigade had overwhelmed the Germans, that the German army had surrendered, but for the most part she and Jan monitored clashes in the secret dailies, weeklies, and news sheets printed throughout the war to keep partisans informed. The editors also sent copies to the Gestapo HQ "just to facilitate your research, [and] to let you know what we think of you. . . ."

German soldiers often came to shoot the flocks of crows that filled the sky like ash before settling in trees. When the soldiers left, Antonina stole out and gathered up the corpses, cleaned and cooked them, making a pâté diners assumed to be pheasant, a Polish delicacy. Once, when ladies praised the rich preserves, Antonina laughed to herself: "Why spoil their appetite with mere details of zoological naming?"

The villa's emotional climate ran to extremes, waves of relaxation followed by a froth of anxiety as people juggled the pastoral pleasantries of one moment with the depressing news of the next. When life sparkled with conversation and piano music, Antonina dodged the war for a spell and even felt delight, especially on foggy mornings when the downtown vanished and she could fancy herself in another land or era.

For that, Antonina told her diary, she was grateful, since life in the lampshade store on Kapucyńska Street had held a steady drizzle of sadness.

Members of the Underground frequently passed through, and sometimes twelve-to-seventeen-year-olds from the Boy Scouts and Girl Scouts. Prominent before the war, youth groups were outlawed during Nazi occupation, but under the aegis of the Home Army they aided the Resistance as soldiers, couriers, social workers, firefighters, ambulance drivers, and saboteurs. Younger scouts enacted minor sabotage like scrawling "Poland will win!" or "Hitler is a Dogcatcher!" (a play on his name) on the walls, a shootable offense, and they became secret letter carriers, while older ones went as far as assassinating Nazi officials and rescuing prisoners from the Gestapo. All helped around the villa by splitting wood, hauling coal, and keeping a fire in the furnace. Some delivered the garden's potatoes and other vegetables to Underground hide-outs, using bicycle rickshaws, a popular vehicle during occupation when taxicabs disappeared and all the cars belonged to Germans.

Inevitably, Ryś overheard scouts whispering alluring secrets and found it frustrating not to join in, when everyone else had exciting, cloak-and-dagger jobs to do. Almost from birth, he'd been schooled in the ambient dangers and told they were real, not pretend or story. Warned not to breathe a word of the Guests to anyone, ever, no matter whom, he knew that if he slipped up, he, his parents, and everyone in the house would be murdered. What a heavy burden for a small child! Full of intrigue and exciting as his world became, with a hodgepodge of eccentric people and dramas, he dared not tell a soul. Small wonder he grew more anxious and worried each day, a fate Antonina lamented in her memoirs, but what could she do when all the adults were anxious and worried, too? Inevitably, Ryś became his own worst nightmare. If a Guest's name or an Underground secret tumbled out while playing, his mother and father would be shot, and even if he survived, he'd be all alone, and it would be *his* fault. Since he couldn't trust himself, avoiding strangers, especially other children, made the most sense. Antonina noted that he didn't even try to make friends at school, but hurried home instead to play with Moryś the pig, whom he could talk with as much as he pleased, and who would never betray him.

Moryś liked to play what they called his "scaredy game," in which he pretended to be scared by some little sound—Ryś closing a book or moving something on a table—and bolted away, hooves skidding on the wooden floor. A few seconds later he would grunt happily next to Ryś's chair, ready for another pretend fright and escape.

Much as Antonina might have wished a normal childhood for Ryś, events had already rusted that possibility and daily life kept corroding. One evening, German soldiers noticed Ryś and Moryś playing in the garden and strolled over to investigate; not fearing humans, Moryś trotted right up to them for a snort and a scratch. Then, as Ryś watched in horror, they dragged Moryś off squealing to be butchered. Shattered, Ryś cried inconsolably for days, and for months he refused to enter the garden, even to pick greens for the rabbits, chickens, and turkeys. In time, he risked the garden world again, but never with the same jubilant nonchalance.

Chapter 15
1941

The pig farm survived only until midwinter, because even in the centrally heated zoo buildings that once housed elephants and hippopotamuses, animals still needed warm bedding. Perversely, it seemed, the "director of slaughterhouses," who funded the zoo, met amiably with Jan and listened to his appeals, but denied him the money to purchase straw.

"This makes no sense at all," he told Antonina afterward. "I cannot believe his idiocy!" Antonina was surprised, because with food scarce, pigs became trotting gold, and how much did straw cost?

"I tried everything I could think of to change his mind," Jan told her. "I don't get it. He has always been our friend."

Antonina declared: "He's a lazy, stubborn fool!"

As the nights crackled with cold and frost feathered the windowpanes, winds knifed through the rinds of wooden buildings and slit life from the piglets. Then an epidemic of dysentery followed, killing much of the remaining herd, and the director of slaughterhouses shut down the pig farm. Infuriating in principle, and depriving the villa of meat, this also thwarted Jan's trips to the Ghetto, supposedly for scraps. Months passed before he learned the truth: in cahoots with another low-level official, the director of slaughterhouses had conspired to rent the zoo to a German herbal plant company.

One day in March, a gang of workmen arrived at the zoo with saws and axes and began dismembering trees, hacking down flower beds, decorative shrubs, and cherished rose-bushes at the entrance gate. The Żabińskis tried screaming, pleading, bribing, threatening, but it was no use. Apparently Nazi orders demanded the uprooting of the zoo, flower and weed alike, because after all, these were only Slav plants, best used as fertilizer for healthy German botanicals. Immigrants usually do try to re-create some of their homeland (especially the cuisine) when they resettle, however this *Lebensraum* didn't only apply to people, Antonina realized, but also to German animals and plants, and through eugenics the Nazis meant to erase Poland's genes from the planet, rip out its roots, crush its hips and tubers, replace its seeds with their own, just as she had feared a year earlier after Warsaw's surrender. Perhaps they felt that superior soldiers needed superior food, which Nazi biology argued could grow only from "pure" seeds. If Nazism hungered for a private mythology, its own botany and biology, in which plants and animals displayed an ancient lineage undiluted by Asiatic or Mideastern blood, that meant starting clean, replacing thousands of Polish farmers and so-called Polish or Jewish crops and livestock with their German equivalents.

At the weekend, by chance, Danglu Leist, the German president of Warsaw and a devotee of zoos, arrived with his wife and daughter, asking for the old zoo director to give them a tour of the grounds and help them imagine the zoo before the war. As Jan strolled with them he compared the

Warsaw Zoo's microclimates to those of zoos in Berlin, Monaheim, Hamburg, Hagenbeck, and other cities, much to Leist's pleasure. Then Jan led his guests to the destroyed rose garden near the main gate where large beautiful bushes, carelessly dug up, lay broken-caned in a pile as casualties of war. Leist's wife and daughter decried the waste of beauty, and that fueled Leist's anger.

"What is *this*?" he demanded.

"It's not my doing," Jan said calmly, with just the right mix of anguish and outrage. He told them about the ruined pig farm and the German herb company renting the zoo from the director of slaughterhouses.

"How could you let that happen?!" Leist raged at Jan.

"What a horrible pity," his wife lamented. "I love roses so much!"

"Nobody asked *me*." Jan quietly apologized to Leist's wife, implying that, since it wasn't his fault, it must be her husband's feckless doings.

She smacked Leist with a hard stare, and he protested angrily: "I didn't know anything about it!"

Before leaving the zoo, he ordered Jan to appear in his office at 10 A.M. the next morning to meet with Warsaw's Polish vice president, Julian Kulski, who would be forced to explain the scandal. When the three men gathered the following day, it transpired that Kulski knew nothing about the scheme, and President Leist promptly canceled the rental agreement, promised to punish the wrongdoers, and asked Kulski for advice on how best to use the zoo without destroying it. Unlike Leist, Jan knew of Kulski's link to the Underground, and as Kulski proposed a public vegetable garden with individual plots, Jan smiled, impressed by a scheme that served the double purpose of cheaply feeding locals and portraying the Nazis as compassionate rulers. Small plots wouldn't destroy the heart of the zoo, but would increase Kulski's influence. Leist approved, and once more Jan changed his career—from zoo director to ruler of a pig farm to magistrate of garden plots. The job bound Jan to the Warsaw Parks and Gardens Department, and that allowed him a new pass into the Ghetto, this time to inspect its flora and gardens. In truth, precious little vegetation grew in the Ghetto, only a few trees by the church on Leszno Street, and certainly no parks or gardens, but he grabbed any excuse to visit friends "to keep up their spirits and smuggle in food and news."

Early on, Antonina had sometimes joined Jan to visit the famous entomologist Dr. Szymon Tenenbaum, his dentist wife Lonia, and their daughter Irena. As boys, Jan and Szymon attended the same school and became friends who loved crawling around in ditches and peering under rocks, Szymon a bug zealot even then. The scarab-like beetle became his sun god, speciality, and mania. As an adult, he started traveling the world and collecting in his spare time, and by publishing a five-volume study of the beetles of the Balearic Islands, he joined the ranks of leading entomologists. During the school year, he served as principal of a Jewish high school, but he collected many rare specimens in Białowieża during the summer, when bugs thronged and any hollow log might hide a tiny Pompeii. Jan, too, liked beetles, and once conducted a large cockroach study of his own.

Even in the Ghetto, Szymon continued to write articles and collect insects, pinning his quarry in sap-brown wooden display boxes with glass fronts. But when Jews were first ordered into the Ghetto, Szymon worried how to protect his large, valuable collection and asked Jan if he'd hide it in the villa. Luckily, in 1939 when the SS raided the zoo and stole over two hundred valuable books, many of the microscopes, and other equipment, they somehow overlooked Tenenbaum's collection of half a million specimens.

The Żabińskis and Tenenbaums became closer friends during the war, as the catastrophe of everyday life drew them tightly together. War didn't only sunder people, Antonina mused in her memoirs, it could also intensify friendships and spark romances; every handshake opened a door or steered fate. By chance, because of this friendship with the Tenenbaums, they met a man who, unknowingly, helped solidify Jan's link with the Ghetto.

One Sunday morning during the summer of 1941, Antonina watched a limousine stop in front of the villa and a heavyset German civilian emerge. Before he could ring the doorbell, she ran to the piano in the living room and started pounding out the loud, skipping chords of Jacques Offenbach's "Go, go, go to Crete!" from *La Belle Hélène*, as the signal for Guests to slip into their hiding places and be silent. Antonina's choice of composer says much about her personality and the atmosphere in the villa.

A German-French Jew, Jacques Hoffmann was the seventh child of a cantor, Isaac Judah Eberst, who, for some reason, decided one day to assume the name of his birthplace, Offenbach. Isaac had six daughters and two sons, and music animated the whole family's life, with Jacques becoming a cello virtuoso and composer who played in cafés and fashionable salons. Fun-loving and satiric, Jacques couldn't resist a prank, personally or musically, and chafing authority was his favorite pastime—he was so often fined for shenanigans at the solemn Paris Conservatory that some weeks he didn't receive any salary at all. He loved composing popular dances, including a waltz based on a synagogue melody, which scandalized his father. In 1855, he opened his own musical theater "because of the continued impossibility of getting my work produced by anybody else," he said wryly, adding that "the idea of really gay, cheerful, witty music—in short, the idea of music with life in it—was gradually being forgotten."

He wrote enormously popular farces, satires, and operettas which captivated the elite and were sung in the streets of Paris, saucy and rollicking music that mocked pretensions, authority, and the idealizing of antiquity. And he cut a colorful figure himself in pince-nez, side-whiskers, and flamboyant clothes. Part of the reason his music besotted so many is that, as music critic Milton Cross observes, it came during "a period of political repression, censorship, and infringement on personal liberties." As "the secret police penetrated into the private lives of citizens . . . the theater went in for gaiety, levity, tongue-in-cheek mockery."

Bubbling with farce and beautiful melodies, *La Belle Hélène* is a comic opera full of wit and vivacity that tells the tale of beautiful Helen, whose boring husband Menelaus wages war with the Trojans to avenge Helen's abduction. The drama caricatures the rulers, bent on war, questions morality, and celebrates the love of Helen and Paris, who want desperately to escape to a better world. Act I ends with the Pythian Oracle telling Menelaus he must go to Greece, and then the chorus,

Helen, Paris, and most of the cast shooing him away in a madcap, galloping "Go, go, go to Crete!" Its message is subversive, ridiculing the overlords and championing peace and love—the perfect signal for the villa's Helens and Parises. Even better, it was by a Jewish composer at a time when playing Jewish music was a punishable offense.

Jan answered the door.

"Does the ex-director of the zoo live here?" a stranger asked.

Moments later, the man entered the house.

"My name is Ziegler," he said, and introduced himself as director of the Warsaw Ghetto's Labor Bureau, the office which, in theory, found work for the unemployed inside and outside the Ghetto, but in practice organized workgangs, deporting the most skillful to serve in armaments factories like the Krupps steelworks in Essen, but did little to help the vast numbers of hungry, semi-employed, often ill workers created by Nazi rule.

"I am hoping to see the zoo's remarkable insect collection, the one donated by Dr. Szymon Tenenbaum," Ziegler said. Hearing Antonina's buoyant piano-playing, he smiled broadly and added: "What a cheerful atmosphere!"

Jan led him into the living room. "Yes, our home is very musical," Jan said. "We like Offenbach *very much.*"

Grudgingly, it seemed, Ziegler conceded, "Oh, well, Offenbach was a shallow composer. But one has to admit that, on the whole, Jews are a talented people."

Jan and Antonina exchanged anxious glances. How did Ziegler know about the insect collection? Jan later recalled thinking: "Okay, I guess this is it, doomsday."

Seeing their confusion, Ziegler said, "You are surprised. Let me explain. I was authorized by Dr. Tenenbaum to view his insect collection, which apparently you are keeping for him in your house."

Jan and Antonina listened warily. Diagnosing danger had become a craft like defusing live bombs—one tremble of the voice, one error in judgment, and the world would explode. What was Ziegler planning? If he wanted, he could just take the insect collection, no one would stop him, so it was pointless to lie about keeping it for Szymon. They knew they had to answer fast to avoid arousing suspicion.

"Oh, yes," Jan said with achieved casualness, "Dr. Tenenbaum left his collection with us before he moved into the Ghetto. Our building is dry, you see, we have central heating; whereas his collection could so easily be damaged in a wet, cold room."

Ziegler shook his head knowingly. "Yes, I agree," he said, adding that he too was an entomologist, an amateur one, who found insects endlessly fascinating. That was how he came to know Dr. Tenenbaum in the first place; but, as it happened, Lonia Tenenbaum was also his dentist.

"I see Szymon Tenenbaum often," he continued with relish. "Sometimes we take my car and drive to the outskirts of Warsaw, where he finds insects in the culverts and ditches. He's an excellent scientist."

They showed Ziegler to the cellar of the administration building, where shallow rectangular boxes stood upright on the shelves like a matching set of old books, each one bound in varnished brown wood with dovetailed joints, glass covers, small metal latches, and a simple number on each spine instead of a title.

Ziegler pulled one box after another from the shelves and

Diane Ackerman

held them up to the light, where they offered a panorama of Earth's coleoptera: gemlike iridescent green beetles collected in Palestine; metallic blue tiger beetles with tufted legs; red-and-green Neptunides beetles from Uganda that cast a sheen like satin ribbon; slender leopard-spotted beetles from Hungary; *Pyrophorus noctilucus*, a little brown beetle more luminous than a firefly, which glows so brightly that South American natives trap several in a lantern to light a hut or tie a few to their ankles to shine their path at night; featherwings, the smallest known beetles, with wings mere stalks edged by tiny hairs; olive-green male Hercules beetles eight inches long, from Amazonia (where natives wear them as necklaces), each sporting such medieval jousting weapons as a giant sword-shaped horn that curves forward overhead and a smaller notched horn curving upward to meet it; female Hercules beetles, also giant, but hornless with beaded wing cases carpeted with red hairs; Egyptian dung beetles like those incised on death-chamber stones; heavily antlered stag beetles; beetles with long looping antennae that bounce overhead like tram wires or lariats; dimple-shelled, cyanide-blue palmetto beetles, which oil sixty thousand short yellow bristles on the soles of their feet to cling impossibly tight to waxy leaves; palmetto beetle larvae wearing straw hats thatched from their own feces, extruded strand by golden strand from an anal turret; net-winged beetles from Arizona with orange-brown wing covers tipped in black, whose hollow wing veins form lacy ridges and cross-ridges filled with noxious blood it dribbles out to repel attackers; hard-to-catch oval whirligig beetles that stride on surface tension near creek banks and ooze a nasty white sap; shiny brown meloid beetles, known whole as "blister beetles" and powdered as "Spanish fly," brimming with cantharidin, a toxin that in small doses spurs erection and in only slightly larger doses kills (Lucretius is said to have died from cantharidin poison); brown Mexican bean beetles that ooze alkaloid blood from their knee joints to deter attackers; beetles with antennae topped by small combs, knobs, brushes, hooves, fringes, or honey dippers; beetles with faces like toothy Halloween pumpkins; fluorescent beetles the blue of Delft miniatures.

Every large beetle monopolized one ball-tipped pin, but smaller beetles floated above one another, sometimes three to a pin. A white flag at the base of each pin told lineage in blue ink graced by swirling capitals, seraph *f*'s and *d*'s, written small but legibly in a steady, meticulous hand. Clearly, collecting the insects fed only part of Tenenbaum's absorption; he also cherished hours wielding microscope, pen, labels, specimens, tweezers, and display boxes crafted for museum drawers and drawing room walls, like those of his contemporary, Surrealist artist Joseph Cornell. How long had Tenenbaum curved over the minute piety of delicately arranging the beetles' legs, antennae, and mouth parts to advantage? Like Lutz Heck, Tenenbaum went on safaris, returning with beetles mounted like deer heads under glass, but more trophies could be hung on the walls of his lap-sized rooms than in any lodge or zoological museum. The sheer time it took to catalogue, gas, prepare, and pin them humbles the mind.

In one glass aerodrome sat row upon row of bombardier beetles, which can zap an attacker with a jet of scalding chemicals fired from a gun turret at the tip of its abdomen.

Harmless when stored separately, the hypergolic chemicals combine in a special gland to concoct a potion volatile as nerve gas. A master of defense and weaponry, the bombardier swivels its gun turret, aims straight at a foe, and fires a 26-mile-per-hour blast of searing irritants—not in a continuous stream, but as a salvo of minute explosions. Thanks to Charles Darwin's misfortune, Tenenbaum knew the bombardier squirts a burning fluid (Darwin was foolish enough to hold one in his mouth while picking up two other bugs). But its secret chem lab was discovered only long after the war by Thomas Eisner, son of a chemist father (whom Hitler had ordered to extract gold from seawater), and a Jewish mother who painted expressionist canvases. The family fled to Spain, Uruguay, and then the United States, where Thomas became an entomologist and discovered that the bombardier's pulsed jet was oddly similar to the propulsion system Wernher von Braun and Walter Dornberger created for 29,000 German V-1 buzz bombs at Peenemünde. Bombardier beetles fire quietly, but the pulse jets of the V-1, flying at about 3,000 feet, buzzed loudly enough to terrify city dwellers as they raced overhead at 350 miles per hour. Only the pause of the telltale buzz spelled death, because when a rocket reached its target, the engine suddenly quit, and in the following suspenseful silence it plummeted to earth with a 1,870-pound warhead. The British nicknamed them "doodlebugs," coming full circle to the weaponry of bombardier beetles.

The wonder on Ziegler's face as he peered into one sense-stealing box after another erased any doubts Antonina had about his motives, because "when he saw the beautiful beetles and butterflies, he forgot all about the world." Moving from row to row, fondling individual specimens with his eyes, reviewing armed and armored legions, he stood spellbound.

"Wunderbar! Wunderbar!" he kept whispering to himself. "What a collection! So much work went into it!"

At last he returned to the present, the Żabińskis, his real business. His face flushed and he looked uncomfortable as he said:

"Now . . . the doctor asks if you'll visit him. Possibly I can help, but . . ."

Ziegler's words trailed into a dangerous and inviting silence. Though he didn't risk finishing the sentence, Antonina and Jan both knew what he meant, something too delicate to propose. Jan quickly replied how immensely convenient it would be if he could ride with Ziegler to the Ghetto and see Dr. Tenenbaum.

"I need to consult with Tenenbaum right away," he explained in a professional tone, "to inquire how best to prevent the insect cases from molding."

Then, to douse any suspicion, Jan showed Ziegler his official Parks Department pass into the Ghetto, implying that the favor he asked was merely for a ride in Ziegler's limousine, nothing illegal. Still charmed by the exquisite collection he'd viewed, and determined it survive for posterity, Ziegler agreed, and off they drove.

Antonina knew Jan wanted to ride with Ziegler because most Ghetto gates were heavily guarded by German sentry on the outside and Jewish police on the inside. Occasionally the gates opened to allow someone through on official business, but passes were prized and hard come by, usually requiring connections and bribery. By chance, the office building at the corner of Leszno and Żelazna Streets, which housed the Labor Bureau where Ziegler worked, formed part of the infamous Ghetto wall.

Topped with crushed glass or barbed wire and built by unpaid Jewish labor, the ten miles of wall rose to about twenty feet in places and zigzagged, closing off some streets, bisecting others lengthwise, hitting random dead ends. "The creation, existence, and destruction of the Ghetto involved a perverse civic planning," writes Philip Boehm in *Words to Outlive Us: Eyewitness Accounts from the Warsaw Ghetto*,

> as the blueprints of annihilation were mapped onto a real world of schools and playgrounds, churches and synagogues, hospitals, restaurants, hotels, theaters, cafes, and bus stops. These loci of urban life . . . Residential streets changed into sites of executions; hospitals became places for administering death; cemeteries proved to be avenues of life support. . . . Under the German occupation, everyone in Warsaw became a topographer. Jews especially—whether inside the Ghetto or out—needed to know which neighborhoods were "quiet," where a roundup was being conducted, or how to navigate the sewer system to reach the Aryan side.

The outside world could be glimpsed through cracks in the walls, beyond which children played and housewives strolled home laden with provisions. Watching keyhole life thriving beyond the Ghetto became torture, and in an inspired twist, Warsaw's Uprising Museum (opened in 2005) includes a brick wall with reverse views: holes through which visitors can glimpse daily life *inside* the Ghetto, thanks to archival films.

At first there were twenty-two gates, then thirteen, and finally only four—all corral style and menacing, in stark contrast to Warsaw's delicately ornate wrought-iron gates. Bridges crossed Aryan streets instead of water. Some notorious soldiers patrolled the boundaries of the Ghetto, hunting children who dared wedge through holes in the masonry to beg or buy food on the Aryan side. Because only children were small enough to squeeze through, they became a tribe of daring smugglers and traders who risked death daily as their families' breadwinners. Jack Klajman, a tough Ghetto child who survived the war by hustling and smuggling, recalls a vicious German major the children nicknamed Frankenstein:

Frankenstein was a short, bull-legged, creepy-looking man. He loved to hunt, but I suppose he must have become bored with animals and decided that shooting Jewish children was a more enjoyable pastime. The younger the children, the more he enjoyed shooting them.

He guarded the area in a jeep with a mounted machine gun. As children would climb the wall, Frankenstein and a German assistant would zoom in from out of nowhere on their killing machine. The other man always drove so Frankenstein had quick access to his machine gun.

Often, when there were no climbers to kill, he would summon Ghetto kids who just happened to be in his line of sight—a long way from the wall and with no intention of going anywhere. . . . Your life was over. . . .

He would pull out his gun and shoot you in the back of the head.

As quickly as children gouged holes in the wall, the holes were patched, then new ones dug. On rare occasions a child smuggler stole out through a gate by hiding among the legs of a labor crew, or a priest. The Ghetto walls sealed one church inside, All Saints, whose Father Godlewski not only slipped real birth certificates of deceased parishioners to the Underground, but would sometimes smuggle a child out under his long robes.

Avenues of escape did exist for the brave with friends on the other side and money for lodging and bribes, but an outside host or guardian like the Żabińskis was essential, because one needed a hideout, food, a raft of false documents, and, depending on if one lived "on the surface" or "under the surface," different webs of agreed-upon stories. If one lived on the surface and was stopped by police, even with false documents one might be asked for the names of neighbors, family, friends, who would then be telephoned or interviewed.

Five tram lines crossed the Ghetto, pausing at a gate on either side, but when they slowed for sharp curves, people could jump off or toss bags to passengers. The conductor and Polish policeman on board both had to be bribed—the going rate was two zlotys—and one prayed that Polish passengers would stay mum. In the far corners of the Jewish cemetery, located inside the Ghetto, smugglers sometimes scaled the fence and climbed into one of two adjacent Christian cemeteries. Some people volunteered for the work gangs that left and returned to the Ghetto each day, and then bribed a gatekeeper to miscount the number of workers. Many cooperative German and Polish policemen guarded the Ghetto gates, eager for bribes, and some helped for free from sheer decency.

Beneath the Ghetto existed a literal underground—shelters and passageways, some with toilets and electricity—where people had crafted intersecting routes between and under the buildings. These led to other avenues of escape, such as slipping through a chiseled hole in the brick wall, or wading through sewers whose labyrinths ultimately led to manhole covers on the Aryan side (though sewers only reached three or four feet high and bred noxious fumes). Some people escaped by clinging to the underside of horse-drawn garbage carts that regularly visited the Ghetto and whose drivers often smuggled in food or left behind an old horse. Those who had the money could disappear in a private ambulance or in a hearse carrying supposed converts to Christian cemeteries, provided gatekeepers were bribed not to search delivery trucks and wagons. Each escapee required at least half a dozen documents and changed houses 7.5 times, on average, so it's not surprising that between 1942 and 1943 the Underground forged fifty thousand documents.

Because the wall meandered, the front of Ziegler's building was accessible from the Aryan side of the city while its seldom-used back door opened onto the Ghetto. In the next building, victims of typhus were quarantined, and across the street stood a somber three-story brick school used as a children's hospital. Unlike other gates, this one wasn't policed by Wehrmacht, Gestapo, or even Polish policemen, only a doorman charged with opening the gate for clerks; and so it promised Jan a rare, lightly guarded way in and out. But this wasn't the only building with one door on the Aryan side and one on the Ghetto side. A convenient crossroads for Jews and Poles to meet, for instance, was the District Court building on Leszno Street, whose rear door opened onto a narrow passageway leading to Mirowski Place on the Aryan side. People mingled and whispered in its corridors, traded in jewels, met friends, smuggled food, and relayed messages, while ostensibly attending court proceedings. Bribed guards and policemen looked aside as some Jews escaped, especially children, right up until the rezoning of August 1942, which finally declared the courthouse outside the Ghetto limits.

There was also a pharmacy on Długa Street with entrances on both sides of the Ghetto wall, where an obliging "pharmacist would allow anyone through who could state a good reason," and several municipal buildings where, for a few zlotys, guards sometimes allowed people to escape.

As their limousine arrived at Leszno 80, the Labor Bureau, the driver honked the horn, a guard swung open the gate, the car entered the courtyard, and they climbed out. This humdrum building contained a lifesaving office because only Jews with a labor card allowing them to work in Wehrmacht factories in the Ghetto avoided deportation.

Lingering beside the front door, Jan thanked Ziegler elaborately in a loud voice, and, though surprised by his sudden formality, Ziegler politely waited for Jan to finish, while the doorkeeper eyed them intently. Jan stretched out the scene, talking mainly in German sprinkled with Polish words, ultimately asking the by-now-impatient Ziegler about using this entrance in the future if he had any trouble with the insect collection and needed to consult about it. Ziegler told the guard to let Jan enter whenever he wished. After that, both men went in and Ziegler showed Jan the way to his upstairs office, and, while giving him a tour of the building, pointed out another staircase that led to the Ghetto door. Instead of heading straight to the Ghetto to visit Tenenbaum, Jan thought it best to spend a little time schmoozing in the dusty offices and narrow hallways of the Labor Bureau, where he made a point of saying hello to as many people as possible. Then he went back downstairs and, in a commanding voice, asked the guard to open the front gate. Drawing attention to himself as a loud, pompous, self-important official would make an impression, he reasoned, and he wanted the guard to remember him.

Two days later Jan returned, using the same boorish voice to demand the gate open for him, and the guard obliged with a welcoming gesture. This time, Jan went to the rear staircase, left the building through the Ghetto door, and visited several friends, including Tenenbaum, whom he told of the curious events involving Ziegler.

Tenenbaum explained that Ziegler had byzantine dental problems and was Dr. Lonia's continual patient; not only had Ziegler found a superb dentist in her, but all of his complex costly treatments were gratis. (Either she had no choice in the matter or she offered free treatments to gain his goodwill.) They agreed to exploit Ziegler's passion for entomology as long as possible, and discussed Underground matters. Tenenbaum now served as principal of the secret Jewish high school, and though Jan offered to smuggle him out, Tenenbaum refused, believing that he and his family stood a better chance of survival inside the Ghetto.

So Jan befriended Ziegler, visited him at his office, and occasionally went with him into the Ghetto to visit Tenenbaum and talk about insects. After a while, he became known as Ziegler's confederate, someone well in with the Labor Bureau head, which smoothed the path for him through the gate, and he often returned by himself to sneak in food to various friends. Occasionally he gave the gatekeeper small tips, as was customary, but not too much nor too often to arouse suspicion.

At last the day seemed right to use the gate for the purpose Jan had had in mind from the start—this time an elegantly dressed and well-coached man accompanied him. As usual, Jan asked the guard to open the gate, and he and his "colleague" walked to freedom.

Emboldened by that success, Jan helped five others escape before the guard grew suspicious. According to Antonina, the guard said to Jan:

"I know *you*, but who is this other man?"

Jan feigned insult, and "with thunder in his eyes," yelled: "I told you that this man is *with me!*"

The intimidated guard only managed in a weak voice:

"I know that *you* can come and go whenever you want, but I don't know *this* person."

Danger clung to every nuance. One sign of guilt, one wrong word, too much bullying, and the guard might guess more than ego was at stake, closing a precious canal between the Ghetto and the Aryan city. Quickly reaching into his pocket, Jan casually said to the doorkeeper:

"Oh, *this* thing. This man has a permit, of course."

And with that he revealed his own Parks Department pass

to the Ghetto, a yellow permit given only to German citizens, ethnic Germans, and non-Jewish Poles. Since Jan's bona fides weren't in question, he didn't need to produce two cards. The surprised guard fell silent in embarrassment. Then Jan shook the guard's hand good-naturedly, smiled, and said solemnly: "Don't worry, I never break the law."

From then on, Jan had no problem escorting Aryan-looking Jews to freedom, but unfortunately, the guard didn't pose the only threat. Any clerk from the Labor Bureau might chance by when Jan and a so-called colleague passed and give them away. Sneaking fugitives past the German troops stationed on zoo grounds created another problem, but the Żabińskis devised two schemes that worked throughout the war—hiding Guests either in the hollows of the villa or in the old animal cages, sheds, and enclosures.

Blending into the kitchen's glossy white woodwork, a door with a lever handle led downstairs to a long basement of rudimentary rooms. At the far end of one, Jan built an emergency exit in 1939—a ten-foot corridor tunneling directly to the Pheasant House (an aviary with a small central building) that adjoined the kitchen garden—which became an entryway for those sheltering in the villa and a handy route for delivering meals. Jan installed running water and a toilet in the basement, and pipes from the upstairs furnace kept the basement relatively warm. Sounds traveled easily between floorboards, so although the Guests heard voices from above, they lived in whispers.

Another tunnel, this one crouchably low and enclosed by rusty iron ribs, led into the Lions House, and some Guests hid in the attached shed, even though it lay within shouting distance of the German armaments warehouse. Looking like part of a whale skeleton, the tunnel used to protect handlers squiring big cats to and from their cages.

Ziegler visited the zoo several more times to behold the remarkable museum of insects and socialize with the Żabińskis. Sometimes he even brought Tenenbaum along with him, on the pretext that the collection occasionally needed direct supervision from its collector, and then Tenenbaum spent hours in his own private paradise, on his knees in the garden, collecting more insects.

Ziegler appeared at the zoo one day with the Tenenbaums' golden dachshund, Żarka, tucked under his arm.

"Poor dog," he said. "She would have a much better life here in the zoo."

"Of course, she's welcome to stay," Antonina offered.

Dipping a hand into his pocket, Ziegler produced little pieces of sausage for Żarka, then set her down and left, though Żarka ran after him and scratched at the door, finally lying down beside it in the lingering scent of the last human she knew.

In the following days, Antonina often found her there, waiting for her family to reappear and whisk her back to a tournament of familiar shapes and scents. This hurly-burly villa had too many rooms for Żarka, Antonina decided, dark corners, steps, mazes, bustle; despite short curvy legs, Żarka kept pacing, unable to settle, nosing around through a forest of furniture and strangers. After a while, she settled into villa life, but always startled easily. If someone's footsteps or a banging door broke the silence, the dachshund's shiny skin would shake nervously all along its thin body, as if trying to creep away.

When winter charged in with skyscraper snows and fewer smells for dogs to read like newsprint, Ziegler visited once more. Still rosy-cheeked and roly-poly, wearing the same old glasses, he greeted Żarka fondly and she remembered him at once, jumping onto his lap and nosing around in his pockets for ham or sausage. This time Ziegler had no treats for Żarka, and he didn't play with her either, just patted her absentmindedly.

"Tenenbaum died," he said sadly. "Imagine, I was just talking with him two days ago. He told me so many interesting stories. . . . Yesterday he had internal bleeding . . . and that was the end. An ulcer broke in his stomach. . . . Did you know he was very ill?"

They didn't. There was little else to say after that shocking news and the sorrow they shared. Overcome by emotion, Ziegler stood up so fast that Żarka fell off his lap, and he abruptly left.

After Szymon's death, the villa went into prolonged mourning, and Antonina worried if his wife could survive the Ghetto much longer. Jan devised an escape plan, but where would they hide her? Much as they wished the villa to sail safely through the war with human cargo, it could only provide temporary shelter for most people, even the wives of boyhood friends.

Chapter 16

The animal world thrives on ploy and counterploy, from chameleons and lion-fish blending in with their backdrop to the majestic cons of mammals. A rhesus monkey who decides not to tell his troop-mates about the melon he just found doesn't need a "theory of mind" to deceive them, only a history of that lie yielding benefits. If his troop-mates find out, he'll be pummeled, and that lesson may alter his selfish ways. But many animals have little choice about sharing food and instinctively call others to the meal. The great apes (including us) have been staging clever deceits, lying on purpose, sometimes just athletically—as practice or sport—for at least 12 million years. Trained interrogators can read the clues of a higher voice, swollen pupils, less eye contact, more complaining, and also learn what "tells" to try and hide.

As a zoologist, Jan had spent years studying the minutiae of animal behavior—all the fineries of courtship, bluff, threat, appeasement gestures, status displays, and many dialects of love, loyalty, and affection. Extrapolating from their behaviors to those of humans came naturally to such a diligent zoologist, especially strategies of deceit. He could adopt new personas fast, a gift that served his shadow life in the

Underground army and also suited his temperament and training.

Not only the Żabińskis, but all Guests and visitors had to cultivate paranoia and abide by the strict rules of their little fiefdom, which meant Ryś and any other children in the house inhaled varieties of truth. Along with languages, they absorbed the lessons of façade-building, tribal loyalty, self-sacrifice, persuasive lying, and creative deception. How do you concoct apparent normalcy? Everything had to appear unremarkable in the household, even if that meant wholly fictitious routines. *Pretend to be normal.* From whose perspective? Would the prewar routines of a Polish zoo director's family seem normal to a patrolling German soldier? The Germans knew the Poles as a deeply sociable people, often with several generations living in one household, plus visiting relatives and friends. So a certain amount of hubbub made sense, but too many lodgers might arouse suspicion.

The current director of the Warsaw Zoo, Jan Maciej Rembiszewski, who, as a boy, volunteered at Jan's zoo (and told him he planned to be a zookeeper himself when he grew up), remembers Jan as a strict boss, a perfectionist, and Antonina depicts him as a demanding paterfamilias, who couldn't tolerate sloppy work or loose ends. From her, we learn that Jan's motto was: "A good strategy should dictate the right actions. Any action mustn't be impulsive, but analyzed along with *all* its possible outcomes. A solid plan always includes many backups and alternatives."

After Szymon's death, Jan visited his wife, Lonia, with details of an escape plan, and news that friends in the Underground were aligning the right stepping-stones so that, after her brief zoo stay, she could vanish to a safer place in the country, maybe even work again as a dentist.

When Jan and Lonia reached the front gate of the Labor Bureau, he intended to use the same ruse he always did and say she was an Aryan colleague who had accompanied him to see Ziegler, since by now the guard was used to his comings and goings, alone or with colleagues. Just as they arrived at the door and he prepared to shepherd Lonia through, he stopped, dismayed to find the guard missing and a woman—the guard's wife, as it turned out—standing in his place. The offices above bustled with Germans only a yell away. She seemed to recognize him, either because she used to watch from the window of a nearby flat or because her husband had described him and his loutish ways, but Lonia's presence troubled her and she became flustered. Not prepared for exceptions, she refused to open the gate.

"We have been visiting Mr. Ziegler," Jan explained firmly.

She said: "Fine, I will open the gate if Mr. Ziegler comes down and personally authorizes your departure."

Her husband had responded well to browbeating, but Jan hesitated—how would verbal abuse work on this woman? Not well, he decided. Staying in character as the arrogant loudmouth her husband knew, he insisted:

"What are you doing? I come here every day, and your husband knows me very well. Now you're *ordering me* to go back upstairs and pester Mr. Ziegler! It will cost you . . . !"

Wavering a little, still unsure, she watched Jan's face grow hot with anger as he snarled like a man fully capable of retaliation, and at last she quietly opened the gate to let them pass. What happened next jarred both Jan and Lonia: right across the street stood two German policemen, smoking and talking while staring their way.

According to Antonina, Lonia described the scene later in words filled with "terror and racing thoughts":

> I wanted to tell Jan—"Let's run." I wanted to get away from that place. I was hoping they wouldn't stop us! But Jan didn't know how I felt, and instead of running, he stopped and picked up a cigarette butt, perhaps left on the sidewalk by these two policemen. Then very slowly he moved his hand under my arm and we started to go toward Wolska Street. This moment felt as long as a century!

That night, passing by the upstairs bedroom, Antonina happened to see Lonia quietly crying into her pillow, with Żarka's wet nose pressed sympathetically against her cheek. Lonia had watched Szymon die; her daughter had been discovered by the Gestapo in Kraków and shot; only the dachshund survived as family.

After a few weeks, the Underground found her safer lodgings in the country, and as Lonia was saying goodbye, Żarka ran up carrying a leash in her mouth.

"You have to stay behind; we don't have a home yet," Lonia told her.

Antonina noted in her memoirs that she found this scene wincingly sad, and that Lonia survived the war, but not Żarka. One day the dachshund, nosing around the German warehouse, ate some rat poison, and after dragging herself back to the villa, died in Antonina's lap.

Three weeks before the Warsaw Uprising, Jan moved Szymon's insect collection to the safety of the Natural History Museum, and after the war Lonia donated it to the State Zoological Museum, in one of whose satellite buildings 250,000 of the original specimens reside today, in a village about an hour north of Warsaw.

To view Tenenbaum's collection, one turns down a narrow macadam road, past an animal hotel (a new concept borrowed from America), past a Christmas tree farm full of pert rows of spruce, to a wooded dead end occupied by two single-story buildings owned by the Polish Academy of Science. The smaller one contains offices, the other miscellaneous overflow from the Zoological Museum.

Entering that huge attic of a building, one finds a divine clutter of millions of specimens where many oddities scream for attention, from stuffed jaguars, lynxes, and native birds to shelves of glass jars crammed with snakes, frogs, and reptiles. Long wooden cabinets and drawers divide one part of the room into narrow alleys of garaged treasures. Tenenbaum's boxed insects occupy two lockers—twenty boxes per shelf, stored upright like books, five shelves per locker. This represents about half of the full collection, which Jan told a journalist ran to four hundred boxes, and Antonina recalled as eight hundred. According to museum records, "Szymon Tenenbaum's wife donated . . . c. 250,000 specimens after the war." At the moment, the boxes remain intact but the archival plan is to remove the insects and file them with many others according to order, suborder, family, genus, and species—all the bombardier beetles in one locker, all the featherwings in another. What a sad dismantling that would be. Certainly the insects would be easier to study, but not the unique vision and artistry of the collectors, who belong to an exotic suborder of *Homo sapiens sapiens* (the animal that knows and knows it knows).

An insect collection is a silent oasis in the noisy clamor of the world, isolating phenomena so that they can be seen undistractedly. In that sense, what is being collected are not the bugs themselves but the deep attention of the collector. That is also a rarity, a sort of gallery that ripples through the mind and whose real holdings are the perpetuation of wonder in a maelstrom of social and personal distractions. "Collection" is a good word for what happens, because one becomes collected for a spell, gathering up one's curiosity the way rainwater collects. Every glass-faced box holds a sample of a unique collector's high regard, and that's partly why people relish studying them, even if they know all the bug parts by heart.

So it doesn't really matter where the boxes sit, but Szymon would have enjoyed this end-of-the-lane, out-of-the-way place, surrounded by farm fields and dense foliage askitter with insects, tiny beetles abounding, where his golden Żarka could chase birds and moles, a dachshund's prerogative. One often recognizes only in hindsight a coincidence or unlikely object that altered fate. Who would have imagined that a zealous professor's cavalcade of pinned beetles would open the gate from the Ghetto for so many people?

Chapter 17

Ziegler's infatuation with insects differed strikingly from Nazi doctrine. Obsessed with pest control, the Third Reich funded many research projects before and during the war that focused on insecticides, rat poisons, and clever ways to foil wood-eating beetles, clothes moths, termites, and other banes. Himmler had studied agriculture in Munich, and favored such entomologists as Karl Friederichs, who sought ways to stop the spruce sawfly and similar insect pests, while justifying Nazi racist ideology as a form of *ecology*, a "doctrine of blood and soil." From this perspective, killing people in occupied countries and replacing them with Germans served both political and ecological goals, especially if one first planted forests to change the climate, as suggested by Nazi biologist Eugene Fischer.

Seen through an electron microscope (invented in Germany in 1939), a louse looks like a pudgy long-horned devil with bulging eyes and six snaring arms. A military scourge in 1812, the bug vanquished Napoleon's Grande Armée en route to Moscow, a legend only recently confirmed by scientists. "We believe that louse-borne diseases caused much of the death of Napoleon's army," Didier Raoult, of the Université de la

Méditerranée in Marseille, reported in the January 2005 issue of the *Journal of Infectious Diseases*, based on an analysis of tooth pulp from soldiers' remains discovered in 2001 by construction workers in a mass grave near Vilnius, Lithuania. As body lice transmitted the agents of relapsing fever, trench fever, and epidemic typhus, Napoleon's Grand Army dropped from 500,000 to 3,000, mainly through pestilence. Friedrich Prinzing's *Epidemics Resulting from Wars*, published in 1916, tells the same tale, and also points out that more men died from lice-borne diseases in the American Civil War than on its battlefields. By 1944, the Germans had medicine to reduce the severity of typhus, but not a reliable vaccine. Nor did the U.S. military, which could only offer its troops repeated typhus inoculations that lasted just a few months.

Inside the Ghetto, crowded apartment buildings quickly became hovels ravaged by tuberculosis, dysentery, and famine, and typhus plagued the Ghetto with high fever, chills, weakness, pain, headaches, and hallucinations. Typhus, a catchall name given to similar diseases caused by *Rickettsiae* bacteria, derives from the Greek word *typhos*, "smoky" or "hazy," limning the mental blur of the sufferer, who, after a few days, develops a rash that gradually covers the whole body. Since lice spread the disease, jamming people into a Ghetto made epidemic inevitable, and in time typhus grew so rife that, passing on the street, people kept their distance for fear of lice jumping onto them. The few doctors, doling out sympathy and care in the absence of medicine and nutrition, knew recovery depended solely on age and overall health.

This led naturally to the image of virulent, lice-ridden Jews. "Antisemitism is exactly the same as delousing," Himmler told his SS officers on April 24, 1943. "Getting rid of lice is not a question of ideology. It is a matter of cleanliness. . . . We shall soon be deloused. We have only 20,000 lice left and then the matter is finished within the whole of Germany."

As early as January 1941, Warsaw's German Governor Ludwig Fischer reported that he chose the slogan "JEWS—LICE— TYPHUS" to emblazon 3,000 large posters, 7,000 small posters, and 500,000 pamphlets, adding that "the Polish press [under German patronage] and the radio have shared in the distribution of this information. In addition, the children in Polish schools have been warned of the danger every single day."

Once the Nazis recategorized Jews, Gypsies, and Slavs as nonhuman species, the image of themselves as hunters naturally followed, with shooting parties at country manors and mountain resorts that prepared the Nazi elite, through blood sport, for the grander hunt. They had other models to choose from, of course, including knights and doctors, but hunter offered the manly metaphors of angling, hounding, baiting, trapping, gutting, ratting, and so on.

The specter of *contagion* clearly unnerved the Nazis. Posters often caricatured Jews with ratlike faces (rat fleas being the primary carriers of plagues), and this imagery insinuated itself even into the psyche of some Jews, like Marek Edelman, a leader of the Ghetto Uprising, who recalled being en route to an Underground meeting when he was "seized by the wish not to have a face," lest someone recognize and denounce him as a Jew. What's more, he saw himself with

a repugnant, sinister face. The face from the poster "JEWS—LICE—TYPHUS." Whereas everybody else . . . had fair faces. They were handsome, relaxed. They could be relaxed because they were aware of their fairness and beauty.

In the bell-jar politics of Ghetto society, rife with social contrasts, criminals and collaborators thrived while others starved, and an underworld of bribery and racketeering arose. German soldiers regularly dished out violence, stole possessions, and grabbed people for backbreaking and humiliating jobs, until, as one resident of the Ghetto wrote, "when the three horsemen of the Apocalypse summoned by the invader—pestilence, famine, and cold—proved no match for the Jews of the Warsaw Ghetto, the knights of the SS were called to complete the task." According to German figures, they shipped 316,822 people from Warsaw to concentration camps between early 1942 and January 1943. Since they also shot many people in the Ghetto, the real death count rose much higher.

Aided by friends on the Aryan side, tens of thousands of Jews managed to escape from the Ghetto before the war ended, but some famously stayed, such as Kalonymus Kalman Shapira, the Ghetto's Hasidic rabbi. Shapira's hidden sermons and diary, unearthed after the war, reveal a tigerish struggle with faith, a man wedged between his religious teachings and history. How could anyone reconcile the agony of the Holocaust with Hasidism, a dancing religion that teaches love, joy, and celebration? Yet one of his religious duties was healing the suffering of his community (not an easy task given the suffering and with all the trappings of piety outlawed). Some scholars gathered at a shoe repair shop and discussed holy texts as they cut leather and hammered in nails, and *Kiddush ha-Shem*, the principle of service to God, acquired a new definition in the Ghetto, where it became "the struggle to preserve life in the face of destruction." A similar word arose in German—*überleben*—which meant "to prevail and stay alive," a defiant point underscored by its being an intransitive verb.

Shapira's Hasidism included transcendent meditation, training the imagination and channeling the emotions to achieve mystical visions. The ideal way, Shapira taught, was to "witness one's thoughts to correct negative habits and character traits." A thought observed will start to weaken, especially negative thoughts, which he advised students not to enter into but examine dispassionately. If they sat on the bank watching their stream of thoughts flow by, without being swept away by them, they might achieve a form of meditation called *hashkatah*: silencing the conscious mind. He also preached "Sensitization to Holiness," a process of discovering the holiness within oneself. The Hasidic tradition included mindfully attending to everyday life, as eighteenth-century teacher Alexander Susskind taught: "When you eat and drink, you experience enjoyment and pleasure from the food and drink. Arouse yourself every moment to ask in wonder, 'What is this enjoyment and pleasure? What is it that I am tasting?"

The most eloquent rabbi and writer of Hasidic mysticism, Abraham Joshua Heschel, left Warsaw in 1939 to become an important professor at the Jewish Theological Seminary in New York (and in the 1960s, a vocal activist for integration). In prose full of koan-like paradoxes, epigrams, and parallels

("Man is a messenger who forgot the message," "Pagans exalt sacred things, the Prophets extol sacred deeds," "The search of reason ends at the shore of the known," "The stone is broken, but the words are alive," "To be human is to be a problem, and the problem expresses itself in anguish"), he felt "loyal to the presence of the ultimate in the commonplace," and that it's "in doing the finite [we] may perceive the infinite." "I have one talent, " he wrote, "and that is the capacity to be tremendously surprised, surprised at life, at ideas. This is to me the supreme Hasidic imperative: Don't be old. Don't be stale."

Most people know that 30 to 40 percent of the world's Jews were killed during World War II, but not that 80 to 90 percent of the Orthodox community perished, among them many who had kept alive an ancient tradition of mysticism and meditation reaching back to the Old Testament world of the prophets. "In my youth, growing up in a Jewish milieu," Heschel wrote of his childhood in Warsaw, "there was one thing we did not have to look for and that was exaltation. Every moment is great, we were taught, every moment is unique."

The etymology of the Hebrew word for prophet, *navi*, combines three processes: *navach* (to cry out), *nava* (to gush or flow), and *navuv* (to be hollow). The task of this meditation was "to open the heart, to unclog the channel between the infinite and the mortal," and rise into a state of rapture known as *mochin gadlut*, "Great Mind." "There is only one God," Hasidic teacher Avram Davis writes,

by which we mean the Oneness that subsumes all categories. We might call this Oneness the ocean of reality and everything that swims in it [which abides by] the first teaching of the Ten Commandments. [T]here is only one *zot*, thisness. Zot is a feminine word for "this." The word zot is itself one of the names of God—the thisness of what is.

The weak, sick, exhausted, hungry, tortured, and insane all came to Rabbi Shapira for spiritual nourishment, which he combined with leadership and soup kitchens. How did he manage such feats of compassion while staying sane and creative? By stilling the mind and communing with nature:

> One hears the [Teaching's] voice from the world as a whole, from the chirping of the birds, the mooing of the cows, from the voices and tumult of human beings; from all these one hears the voice of God. . . .

All our senses feed the brain, and if it diets mainly on cruelty and suffering, how can it remain healthy? Change that diet, on purpose, train mentally to refocus the mind, and one nourishes the brain. Rabbi Shapira's message was that, even in the Ghetto, common people could temper their suffering in this way, not just monks, ascetics, or rabbis. It's especially poignant that he chose for meditative practice the beauty of nature, because for most people in the Ghetto nature lived only in memory—no parks, birds, or greenery existed in the Ghetto—and they suffered the loss of nature like a phantom-limb pain, an amputation that scrambled the body's rhythms, starved the senses, and made basic ideas about the world impossible for children to fathom. As one Ghetto inhabitant wrote:

In the ghetto, a mother is trying to explain to her child the concept of distance. Distance, she says, "is more than our Leszno Street. It is an open field, and a field is a large area where the grass grows, or ears of corn, and when one is standing in its midst, one does not see its beginning or its end. Distance is so large and open and empty that the sky and the earth meet there.... [Distance is] a continuous journey for many hours and sometimes for days and nights, in a train or a car, and perhaps aboard an airplane.... The railway train breathes and puffs and swallows lots of coal, like the ones pictured in your book, but is real, and the sea is a huge and real bath where the waves rise and fall in an endless game. And these forests are trees, trees like those in Karmelicka Street and Nowolipie, so many trees one cannot count them. They are strong and upright, with crowns of green leaves, and the forest is full of such trees, trees as far as the eye can see and full of leaves and bushes and the song of birds."

Before annihilation comes an exile from Nature, and then only through wonder and transcendence, the Ghetto rabbi taught, may one combat the psychic disintegration of everyday life.

Chapter 18

1941

As summer passed into autumn, flocks of bullfinches, red crossbills, and waxwings began streaming south from Siberia and northern Europe along sky corridors older than the Silk Road, passing overhead in squadron V-formations. Because Poland lies at the intersection of several great flyways—south from Siberia, north from Africa, west from China—autumn laced the air with a stitchery of migrating songbirds and chevrons of blaring geese. Insect-eating birds flew deep into Africa, with the spotted flycatcher, for example, covering thousands of miles and flying nonstop for about sixty hours over the Sahara. Not needing to fly quite so far, great blue herons and other waders settled along the shores of the Mediterranean, Atlantic, Caspian, or Nile. Nomadic birds needn't follow a strict route; during the war, some veered off east and west, avoiding bomb-scented Warsaw entirely, though much of Europe proved equally inhospitable.

At the villa, Guests and visitors migrated in late autumn to warmer rooms or more durable hideouts. The Żabińskis faced their third wartime winter with a stockpile of coal so meager they could warm only the dining room, provided they first drained the water from the radiators and sealed off the staircase and second floor. That divided the house into three climates: subterranean dank, first-floor equator, and polar bedrooms. An old American woodstove borrowed from the Lions House smoked irritably, but they huddled beside it anyway, peering through a small glass door at red-and-blue flames licking chunks of coal and periodically rinsing them with fire. As the chimney piped a hymn of warmth, they enjoyed the wordless magic of conjuring heat indoors on frigid days. Bundled up in fleece and flannel, Jan and Ryś could sleep beneath further layers of blankets and down comforters, then spring from bed and stay warm just long enough to dress for work or school. The kitchen felt like a meat locker, frost embroidered the windows inside and out, and fixing meals, doing dishes or, worse yet, the laundry—any chore that meant dipping her hands in water—tortured Antonina, whose skin chapped until it bled. "Slick-skinned humans just aren't adapted to fierce cold," she mused, except by using their wits, donning the hides of animals, trapping smoky fires.

Each day, after Jan and Ryś had left, she hitched up a sled and pulled a barrel of scraps from the slaughterhouse to the chicken shed, then fed the rabbits hay and carrots from the summer garden. While Ryś attended Underground school several blocks away, Jan worked downtown, in a small lab that inspected and disinfected buildings, a minor job that doled out useful perks: food stamps, a daily meal of meat and soup, a work permit, a little pay, and something priceless to the Underground—legal access to all parts of the city.

Because they hadn't enough fuel to heat the cages, sheds, and three floors of the villa, all the Guests were spirited away to other winter safe houses, either in Warsaw or the suburbs. The Underground hid some Jews on country estates that, instead of being confiscated, stayed in the owners' hands to produce food for German troops. There, an illegal woman could assume the role of governess, maid, nanny, cook, or tailor; and a man work in the fields or at the mill. Others might hide with peasant farmers or as teachers in the communal schools. One such estate, owned by Maurycy Herling-Grudziński, lay only about five miles west of downtown Warsaw, and at one time or another, five hundred or so refugees sheltered there.

Even with the Guests and relatives gone, the wintry villa included two eccentric tenants, and according to Antonina, the first to arrive, Wicek (Vincent), belonged to an aristocratic family of impeccable lineage. "His mother was a member of a famous line of silver rabbits" known as arctic hares, a breed whose young start out glossy black and silver up later on as they pale into adolescence. October's wet gales made Wicek shiver in a garden hutch, so Antonina brought him indoors to the relative warmth of the dining room by day and Ryś's heavily blanketed bed at night. Each morning, while Ryś dressed for school, Wicek slid from between the bedcovers and hopped along the hallway to the stairwell, then carefully descended the narrow steps and nosed open the wooden divider to scurry into the dining room, where he nestled beside the stove's glass door. There he flattened his long ears against his back for added warmth, and stretched one rear leg straight out while tucking the other three in tight. Naturally gifted with amber eyes outlined in black like Egyptian hieroglyphs, three layers of fur, large snowshoe feet, and extra long incisors for gnawing moss and lichen, he quickly developed habits and tastes unknown to rabbit culture and a bizarre griffin-like personality.

At first, whenever Ryś sat down to dinner, Wicek draped himself along Ryś's foot like one furry black slipper, instinctively crouching as hares do in arctic windstorms. Then, as Wicek grew large and muscular, he bounced around the house like hard rubber, and at meals hopped from the floor straight onto Ryś's lap, thrust his front paws onto the table, and grabbed Ryś's food. Naturally vegetarian, arctic hares may resort to tree bark and pinecones at times, but Wicek preferred stealing a horse cutlet or slice of beef, and bouncing away to devour it in a shadowy corner. According to Antonina, he'd zoom into the kitchen whenever he heard the thud of her meat-tenderizing hammer, hop onto a stool, leap from there to the table and snatch a slice of raw meat, then dart away with his trophy and feast like a small panther.

During the holidays, when a friend sent the Żabińskis a gift of kielbasa, Wicek became a razor-toothed pest, begging for scraps and mugging anyone he found eating sausage. In time, he also discovered cold cuts hidden atop a piano in Jan's office next to the kitchen. In theory, the piano's slippery legs deterred hungry mice; not so, hungry hares. With all his pilfering, Wicek quickly grew into a fat, furry thug, and whenever they left the house, they jailed him behind a corner cupboard, since he'd begun eating their clothes. One day he chewed the collar of Jan's jacket hanging on a chair in the bedroom; another, he scalloped a felt hat and hemmed a visitor's coat. They joked about his being an attack rabbit, but in a more solemn mood, Antonina wrote that wherever she turned in the human or animal world, she found "shocking and unpredictable behavior."

When a sickly male chicken joined the household, Antonina nursed it back to health and Ryś claimed it as another pet, naming it Kuba (Jacob). In prewar days, the villa had harbored more exotic animals, including a frisky pair of baby otters, but the Żabińskis continued their tradition of people and animals coexisting under one roof, over and over welcoming stray animals into their lives and an already stressed household. Zookeepers by disposition, not fate, even in wartime with food scarce, they needed to remain among animals for life to feel true and for Jan to continue his research in animal psychology. According to Jan, "The personality of animals will develop according to how you raise, train, educate them—you can't generalize about them. Just like people who own dogs and cats will tell you, no two are exactly alike. Who knew that a rabbit could learn to kiss a human, open doors, or give us reminders about dinnertime?"

Wicek's personality intrigued Antonina, too, who declared him "insolent," preternaturally cunning, and even scary at times. A kissing, predatory, carnivorous rabbit—it was the stuff of fairy tales and a good subject for one of her children's books. She kept tabs on his escapades, watching him crouch in wait, ears alert as radar dishes, tracking every noise, straining to decipher sounds.

The indoor zoo created a diverting circus of rituals, odors, and noises, with the bonus of play and laughter, a tonic for everyone, especially Ryś. Animals helped distract him from the war, Antonina thought, so feathered or four-legged, clawed or hoofed, reeking of badger musk or scentless as a newborn fawn, in time all entered his zoo within the villa's menagerie within the old Warsaw Zoo: a matryoshka doll of zoos.

In the villa, some of Antonina's clan sprayed table legs and chairs, some shredded and gnawed or leapt onto the furniture, but she enjoyed them as specially exempt children or wards. House rules decreed that Ryś looked after pets, as a mini-zookeeper who tended a small fiefdom of gnomes even needier than he. This kept Ryś busy with important chores, ones he could master, at a time when everyone else seemed to have grown-up secrets and responsibilities.

There was no way so young a child could comprehend the network of social contacts, payoffs, barter, reciprocal altruism, petty bribes, black market, hush money, and sheer idealism of wartime Warsaw. A house "under a crazy star" helped everyone forget the crazier world for minutes, sometimes hours, at a time, by serving up the moment as a flowing chain of sensations, gusts of play, focused chores, chiming voices. The rapt brain-state of living from moment to moment arises naturally in times of danger and uncertainty, but it's also a rhythm of remedy which Antonina cultivated for herself and her family. One of the most remarkable things about Antonina was her determination to include play, animals, wonder, curiosity, marvel, and a wide blaze of innocence in a household where all dodged the ambient dangers, horrors, and uncertainties. That takes a special stripe of bravery rarely valued in wartime.

While Rabbi Shapira preached meditating on beauty, holiness, and nature as a way to transcend suffering and stay sane, Antonina was filling the villa with the innocent distractions of muskrat, rooster, hare, dogs, eagle, hamster, cats, and baby foxes, which drew people into a timeless natural world both habitual and novel. Paying attention to the villa's unique ecosystem and routines, they could rest awhile as the needs and rhythms of different species mingled. The zoo vistas still offered trees, birds, and garden; sweet linden blooms still hung like pomanders; and, after dark, piano music capped the day.

This sensory blend grew more vital as dozens more Guests arrived with horrific tales of Nazi brutality, and the Żabińskis embraced them, drawing support from "clandestine groups and contacts, some very strange indeed," as Irena Sendler (code name "Jolanta") described it. A Christian doctor's daughter with many Jewish friends, she reconfigured her job at the Social Welfare Department, recruited ten like-minded others, and began issuing false documents with forged signatures. She also wangled a legal pass into the Ghetto via a "sanitary-epidemiological station," supposedly to deal with infectious diseases. In truth the social workers "smuggled in food, medicine, clothing, and money, while freeing as many people as possible, particularly children." That meant first persuading parents to give her their children, then finding ways to smuggle the little ones out—in body bags, boxes, coffins, through the old courthouse or All Saints church—finally placing them with Catholic families or in orphanages. A jar she buried in a garden held lists of the children's real names, so that after the war they might be reunited with family. Nuns often hid children in orphanages in or near Warsaw, with some specializing in hard-to-place Semitic-looking boys, whose heads and faces were bandaged, as if they'd been wounded.

The Żabińskis received word, by telephone or messenger, to expect a Guest for a brief stay, and Irena often visited them in person, with news, just to talk, or to hide when her office fell under surveillance. Later, captured by the Gestapo and brutally tortured in Pawiak Prison, Irena escaped with the Underground's help; she became one of the zoo's favorite Guests.

The Polish government-in-exile, based in London, staffed a radio station and planned missions, borrowing British planes, agents, and resources. Smuggling in cash strapped to parachutists whose money belts held as much as $100,000 and the addresses of recipients in code, Polish agents known as *cichociemni* (pronounced *cheekoh-chemnee*), "the dark and silent ones," also packed weapons, weapon-making kits, and plans. According to one *cichociemni*'s account, to keep dispersal to a minimum his group jumped from 300 feet and aimed for "a cross of red and white flowers impudently alight in a large clearing." Whooshing between pine trees, he landed on his feet and was met by a helmeted man who quickly exchanged password and handshake. Then rural youths appeared to claim the boxes and gather up the parachutes, from which women would sew blouses and underwear. After delivering an encrypted message from the commander-in-chief to the commander of the Home Army, he swallowed the regulation dose of caffeine-laced Excedrin to stay alert and put a cyanide pill in a special pocket of his trousers, before being led to a schoolhouse where a zaftig headmistress fed him a bacon and tomato omelette, sending him on his way at dawn. Some of the jumpers joined local units and many fought in the Warsaw Uprising of 1944. Of 365 couriers, 11 died; 63 aircraft were shot down; and only about half of the 858 drops were successful. But they supplied a tireless Underground, described by ally and enemy alike as the best organized in Europe, and it needed to be, since the Third Reich had singled out the Poles for special punishment.

By now, Jan delved deeper into Underground work and taught general biology and parasitology at the Faculty of Pharmacy and Dental Medicine of Warsaw's "flying university." Classes were small and the meeting rooms nomadic, to avoid discovery, floating from one edge of Warsaw to the other, in private apartments, technical schools, churches, businesses, and monasteries, inside the Ghetto and outside. It issued primary school, bachelor, and graduate degrees in medicine and other professions, despite the lack of libraries, laboratories, and classrooms. A certain sad irony (or perhaps it was optimism) prompted the Ghetto doctors, who could only comfort those dying patients whom a little food and medicine would have cured, to teach cutting-edge medicine to a future generation of doctors. At the outbreak of the war, thinking to decapitate the country, the Nazis had rounded up and shot most of the Polish intelligentsia, then outlawed education and the press, a strategy that boomeranged because it not only made learning subversively appealing, it also freed the surviving intellectuals to focus their brainpower on feats of resistance and sabotage. Widely read clandestine newspapers circulated in and out of the Ghetto, where they were sometimes stacked in Jewish toilets (which Germans scrupulously avoided). In this time of blatant deprivation, libraries, colleges, theater, and concerts flourished, even secret All-Warsaw soccer championships.

By the spring of 1942, a stream of Guests began arriving at the zoo once more, hiding in cages, sheds, and closets, where they tried to forge daily routines while living in a state of contained panic. Versed in the layout of the house, surely they joked about the clunkiness of so-and-so's footsteps, children running, hoof and paw skitterings, door slammings, phone ringing, and the occasional banshee screeching of quarreling pets. At least, in a radio era, they'd grown used to gathering news by ear and adding mental images.

Antonina worried about her friend, sculptor Magdalena Gross, whose life and art had derailed with the bombing of the zoo, which wasn't just her open-air workshop but her compass, in both senses, an imaginative realm for her work and a direction for her life. Antonina wrote in her diary of Gross's rapture, how the animals absorbed her until she lost herself in their quiddities for hours, oblivious to zoogoers who stood quietly watching. Jan, a lifelong fan of what he called "the plastic arts," also admired her work enormously.

Small sculpture her specialty, Gross had captured two dozen or so animals, lifelike and witty, on the brink of a familiar motion or with distinctly human traits: A camel with its head laid back on one hump, legs splayed, caught mid-stretch. A young llama with perked-up ears spying something edible. A wary Japanese goose pointing a sharp beak skyward while eyeing the viewer, like "a beautiful but brainless woman," Gross had explained. A flamingo, mid-Chaplinesque walk, its right heel lifted. A macho pheasant showing off for his harem. An exotic hen hunkering down and trotting fast, "like a shopper thinking only about how to buy some herrings." A deer craning its head backwards when startled by a sound.

A bright-eyed heron with long, solid beak, curvaceous shoulders, and chin plunged deep into a large fluffed-up chest—which Magdalena identified as herself. A tall marabou with head sunk deep between its shoulders. An elk sniffing the air for a whiff of a mate. A feisty rooster, ready for trouble, rolling a wild eye.

Gross sought the innuendos of flesh unique to each animal: how it angled hips and shoulders to balance, threatened rivals, showed emotion. She relished tiny flexions, angling her own arms and legs to understand the rigging of her models' muscles and bones. Jan, who served as Magdalena Gross's advisor, was fascinated by the core design of animals, their center of gravity and geometry—how, for example, a bird balances its low smooth mass on two twiglike legs, while a mammal's richer core of shapes and textures requires the props of four thick legs. With his college studies in agricultural engineering, zoology, and fine arts, he may well have been influenced by Darcy Wentworth Thompson's charming classic, *On Growth and Form* (1917), a study of biological engineering, which considers such motifs as the architecture of the spine or the pelvis evolving bone-wings to spare the torso pain. She spent months crafting a sculpture. To select from a repertoire of moves one pose that might embody it—that took time and a kind of infatuation, an ecstasy of imagining Gross loved. The joy shows in her sculptures.

Antonina often praised her artistry and mused how Magdalena figured in the long saga of humans depicting animals in art, stretching back to the Paleolithic age, when by the light of firebrands, humans drew buffalo, horses, reindeer, antelopes, and mammoths on cave walls. They weren't exactly *drawn*; sometimes pigments were carefully blown onto the wall (the laser-perfect replica cave at Lascaux today was decorated using that technique). Animal fetishes carved from antler and stone joined the reliquary, either for worship or for use by hunters in sacred cave ceremonies. Bulging from the natural contours of limestone walls, the animals galloped through initiation rites, in flickering darkness where one could easily confuse heartbeats and hoofbeats.

At the beginning of the twentieth century, and between the wars in the heydey of Dadaism and Surrealism (neither of which was an *ism* as much as an idea about the role of art in life and life as art), animal sculpture flourished in Polish art, and continued during and after World War II. In Antonina's eyes, Magdalena joined the fluent tradition of magical animals adorning the art of ancient Babylonia, Assyria, Egypt, the Far East, Mexico, Peru, India, and Poland.

Magdalena would first model in clay before fixing a design in bronze, and during this soft, forgiving stage, she often asked Jan to critique the anatomical details of her work, though he reported to Antonina that she rarely erred. Each sculpture took many months to finish, and Magdalena averaged only one bronze a year, because she studied every flake and fiber of her model, dickering with design, and it was hard to let the clay mannequin rest. Once, when someone asked her if she liked her finished handiwork, she said: "I'll answer your question in three years." She cast only two endangered animals—the European elk and bison—devoting two years to the latter, a special gift for Jan. Of course, the zoo animals wouldn't pose—they often took wing, toddled off, or hid from her—and wild animals reserve making eye contact for the rough occasions of eating, mating, or dueling. Vigorously minding them calmed her, which in turn calmed them, and in time they allowed her to stare for longer spells.

Famous as Gross was (her *Bison* and *Bee-Eater* took gold medals at the 1937 International Art Exhibition in Paris), Antonina reckoned her a surprisingly modest woman, endearingly optimistic, and simply besotted with animals and art. Antonina recalled how Gross charmed her models, their patrons, and guards: "Everyone welcomed the sight of this sunny little 'Mrs. Madzia,' with her dark smiling eyes, molding clay with delicacy and gusto."

When Jews had been ordered into the Ghetto, Gross refused, by no means an easier fate, because those who lived on the surface had to disguise themselves as Aryans and keep up the masquerade at all times, cultivating Polish street language and a plausible accent. Estimates vary, but the most reliable, from Adolf Berman (who aided them and kept good records), found 15,000 to 20,000 people still in hiding as late as 1944, and he assumed the number had been much higher. In *Secret City*, a study of the Jews who, at one time or another, lived on the Aryan side, Gunnar Paulsson puts the figure closer to 28,000. As he rightly says, with figures that high, we're really talking about an embedded city of fugitives, complete with its own criminal element (scores of blackmailers, extortionists, thieves, corrupt policemen, and greedy landlords), social workers, cultural life, publications, favorite cafés, and lingo. Jews in hiding were known as *cats*, their hiding places *melinas* (from the Polish for a "den of thieves"), and if a *melina* was discovered, one referred to it as *burnt*. "Consisting of 28,000 Jews, perhaps 70,000–90,000 people who were helping them, and 3,000–4,000 *szmalcowniks* [blackmailers, from the Polish word for lard] and other harmful individuals," Paulsson writes, "[this] population numbered more than 100,000, probably exceeding the size of the Polish Underground in Warsaw, which fielded 70,000 fighters in 1944."

The smallest oversight could give a *cat* away—not knowing the price of a tram ticket, say, or appearing too aloof, not receiving enough letters or visitors, not taking part in the typical social life of a housing block, like this one described by Alicja Kaczyńska:

> Tenants visited each other . . . sharing news about the political situation, often playing bridge. . . . When returning home in the evening . . . I would stop at the little altar in the gateway of our building. The whole of Warsaw had such altars in its gateways, and the whole of Warsaw sang: "Listen, Jesus, how your people plead/ Listen, listen, and intercede." The tenants of our building gathered at these prayers. . . .

Paulsson tells of "Helena Szereszewska's daughter, Marysia, who considered herself completely assimilated and moved about freely," and who "once saw some lemons (almost unobtainable in wartime) on a market stall. Out of curiosity, she asked the price, and when the stall-keeper named the astronomical sum she exclaimed 'Jezu, Maria!' as a Polish Catholic would. The stall-keeper replied slyly: 'You've known them such a short time, missy, and you're already on a first-name basis!' "

Lodging with an old woman, Gross delivered tortes and

pastries for several bakeries, which paid her just enough to survive, and she risked leaving the apartment to meet friends at a *cat*-friendly coffeehouse. Jews in hiding sometimes met at a café at 24 Miodowa Street, or at another on Sewerynów Street, where they could dine at "the Catholic Community Centre of St. Joseph, which had a well-patronized restaurant. The fact that it was in a quiet side street and the service by the nuns was so pleasant attracted many Jews to the place. . . . It was known to nearly all the Jews hidden in Warsaw, and offered an hour's respite from the cruel outside."

Whenever Gross left home, there was always the chance of being recognized and denounced, but in an atmosphere of daily street executions and house searches, Antonina worried when she heard a rumor that Nazis had begun combing through the apartment houses in Magdalena's neighborhood, at odd hours, raiding attics and basements to roust out hidden Jews.

Chapter 19

Antonina stood in the kitchen, kneading bread dough, a daily ritual, when she heard Ryś's excited voice at the back door:

"Hurry up! Starling! Come here!"

Apparently, her son had another new animal friend, and she liked his choice of species. Starlings had always charmed her with their "long, dark beaks, springy hop, and cheerful cackles," and she enjoyed watching them pogo-hop on the ground and dig for worms, tail and head nimbly twitching. The feast of the starlings always foretold winter's end and "the earth softening up its belly for spring." Flocks of starlings form wonderful shapes as they circle the sky—troika reins, kidney beans, cone shells. Turning as one unit, for an eye blink they vanish, then suddenly reappear a moment later like a shake of black pepper. Bouncing and fluttering on the ground, they reminded Antonina of "feathery jesters," she noted in her memoirs, and it pleased her to think of Ryś catching and befriending one. Standing at the sink, hands in gummy dough, she called over her shoulder that she was too sticky to greet his new treasure, but would later. Right then the kitchen door sprang open and she suddenly understood the real meaning of Ryś's words. There stood Magdalena

Gross, wearing an old summer coat and a pair of tattered shoes.

All the Guests and friends in hiding had secret animal names, and Magdalena's was "Starling," in part because of Antonina's fondness for the bird, but also because she pictured her "flying from nest to nest" to avoid capture, as one *melina* after another became *burnt*. Passersby wouldn't be surprised to hear animals mentioned at the zoo, and one gets a sense that it also just *felt right* to Jan and Antonina, that naming the usual animals helped them restore a little normalcy to their lives.

In the topsy-turvy alleyways of occupied Poland, the fame Magdalena had enjoyed before the war now endangered her. What if someone from her past spotted her and, from good or bad motives, told of her whereabouts? Rumor has long ears, and as an old Gypsy saying goes, *Fear has big eyes*. With Magdalena on board, the other Guests had to be doubly careful, and Magdalena dared not show her face, so familiar in some Polish circles. "Madzia's usually happy eyes became a little sad now," Antonina wrote in her diary. Antonina and Jan sometimes also called her "Madzia," an affectionate nickname from the softened form of Magda—as the hard formal *g* becomes a soft *j* sound, it yields to convey tender emotions. "She missed the freedom and exciting lifestyle she had before the war," which included a large circle of friends in the arts. In 1934, for example, Magdalena had helped Bruno Schulz, a Chagall-like painter and author of prose phantasmagoria, find a publisher for his first book, *Sklepy Cynamonowe* (translated as *Cinnamon Shops*), a collection of short stories about his eccentric family. She put Schulz's manuscript into the

Antonina and Jan feeding an injured bird.

Badger taking Ryszard for a walk.

*Family portrait.
From left to right:
Jan's mother,
Antonina, and Jan
holding Ryszard.*

*Teresa holding Badger
after the war.*

The Pheasant House in 1929. During the war, a secret underground passage led from the Pheasant House to the villa. People were also smuggled in through the Lion's House.

A 1937 postcard of Tuzinka, the twelfth elephant ever born in captivity, hence her name from the Polish for "dozen."

Polar bears at the zoo, 1938.

At one point, Antonina hand-raised hyena pups indoors. Orphaned or injured animals quickly became a normal part of the household, which included hamster, piglet, baby badger, arctic hare, muskrat, cockatoo, lynxes, and many other animal companions, as well as hundreds of hidden Jews.

Jan holding a lynx.

Przywalski horses, a rarity in prewar zoos. Jan was understandably proud of the foal.

A resident of the Warsaw Zoo with two tall friends, sometime in the 1930s.

A drawing of the aurochs, one of the extinct animals the Nazis hoped to re-create so that they would have pure Aryan game to hunt after the war.

Jan and Antonina's villa today, seen from the rear.

The postcard Jan sent to Antonina from the POW camp during the war. Although he couldn't risk writing anything, he conveyed his condition and mood in this self-portrait.

Polish horses descended from the extinct tarpan, another nearly mythic animal the Nazis hoped to re-create. Tarpans, aurochsen, and European bison are the animals Neolithic hunters once painted in ochre on cave walls.

The Royal Hunting Lodge in Białowieża, once the ornate retreat of kings and tsars.

hands of another friend, novelist Zofia Nałkowska, who declared it innovative and brilliant, and guided it through publication.

Hiding indoors by day, Magdalena couldn't roam the zoo to find models, so she decided to sculpt Ryś.

"He's a lynx," she joked. "I should have good results with this sculpture!"

One day, as Antonina was kneading dough for bread, Magdalena said: "Now I can help *you*. I learned how to bake delicious croissants. I may not be able to sculpt in clay now, but I can still sculpt in flour!" With that she plunged a palm into a big bowl of dough, sending up a small white cloud.

"It's terrible that such a gifted artist has to work in the kitchen!" Antonina lamented.

"It's only a temporary situation," Magdalena assured her, gently elbowing her away from the bowl and kneading the dough with powerful hands.

"Some might say that a woman as little as I am couldn't be a good baker. Well! Sculptors develop enormous strength!"

Muscling into clay had given her powerful shoulders and hands annealed by her trade. In her circle, which included Rachel Auerbach and Yiddish poet Deborah Vogel, among others, what Bruno Schulz called the "unique mystical consistency" of matter really mattered, as did the hands that handled it. This was a topic their set often discussed in long, thoughtful, literary letters crafted partly as an art form. Few have survived, but, fortunately, Schulz recruited many of his own for short stories.

In Paris, before the war, Magdalena would surely have studied Rodin's vigorous sculptures of hands in the Rodin

Museum, a small music box of a building surrounded by rosebushes and brawny sculptures. She was justly proud of the way strong, agile hands cradle newborns, build cities, plant vegetables, caress loved ones, teach our eyes the shape of things—how round swells, how sand grits—bridge lonely hearts, connect us to the world, map the difference between self and other, fasten onto beauty, pledge loyalty, cajole food from grain, and so much more.

Magdalena seasoned the villa with "loads of sunshine, energy, and a great spirit," Antonina wrote, "which she never lost, even during terrible crises, and she faced horrendous ones in her life. No one ever saw her being depressed." Antonina sometimes wondered how on earth they'd lived without her until then, because she'd become such a robust part of their clan, sharing their lives, everyday concerns, hardships, and insecurities, helping with house chores, and whenever they had too many Guests, giving up her bed and sleeping atop a large trunk for flour, or on two armchairs pushed together. "Like her nickname, Starling, she whistled at hardship, when many in her situation would have succumbed to despair," Antonina recalled in her memoirs. Whenever the household expected a visit from a stranger, Magdalena would hide, and if the visitor seemed dangerous or, worse yet, wanted to go upstairs for some reason, Antonina would alert her with the usual alarm of piano notes or, when that wasn't convenient, a sudden outburst of song. She regarded Magdalena as "a bit of a rascal" and a rousing chorus of Offenbach's "Go, go, go to Crete!" the perfect getaway tune for someone that prankish and high-spirited.

Whenever Magdalena heard that music, she dashed to a

hiding place, which, depending on her mood, might be the attic, a bathroom, or one of the deep walk-in closets. As she confided to Antonina, she usually did so laughing quietly about the absurdity of the situation.

"I wonder," she sometimes joked, "how I'll feel about this music when the war is over! What if it happens to be playing on the radio? Will I dash for cover? Will I even be able to *stand* this song of Menelaus going to Crete?"

Once its sprightly melody had been a favorite of hers, but war plays havoc with sensory memories as the sheer intensity of each moment, the roiling adrenaline and fast pulse, drive memories in deeper, embed every small detail, and make events unforgettable. While that can strengthen friendship or love, it can also taint sensory treasures like music. By associating any tune with danger, one never again hears it without adrenaline pounding as memory hits consciousness followed by a jolt of fear. She was right to wonder. As she said, "It's a terrific way to ruin great music."

Chapter 20

The snowy autumn of 1942 hit the zoo with a special fury, as winds lashed wooden buildings until they moaned and whisked snowbanks into sparkling soufflés. Bombing early in the war ripped up the zoo grounds, scrambling its landmarks, then snow fell heavily, hiding a bevy of new ruts, downed fences, twisted macadam, jagged fingers of metal. Below the deceptively soft snowscape, metal basilisks lurked everywhere, confining people to a maze of shoveled walks and well-trodden pastures.

Antonina's range shrank even more because she was crippled by what sounds like phlebitis (she offers few clues), a painful infection in the leg veins that made walking agonizing and meant bed rest from fall of 1942 to spring of 1943. An unusually active thirty-four-year-old, she hated being confined to her bedroom, in heavy clothes, muffled under strata of blankets and comforters ("I felt so embarrassed and useless," she moaned in ink), when there was a large household to manage. She was the top matryoshka, after all, and not just symbolically, since she was also pregnant. It's hard to know if blood clots did form in her legs—from pregnancy, smoking, varicose veins, heredity? Certainly not inactivity or obesity.

But phlebitis can be dangerous; in its severest form, deep vein thrombosis, a blood clot travels to the heart or lungs, causing death. Even mild phlebitis, or possibly rheumatoid arthritis (an inflammation of the joints), creates red swollen legs and requires bed rest, so, having no alternative, she held court in her bedroom, with family, friends, and staff alike paying calls.

In June of 1942, the Polish Underground received a letter, written in code, telling of an extermination camp at Treblinka, a town not far from Warsaw. Here's part of its warning:

> Uncle is planning (God preserve us) to hold a wedding for his children at your place, too (God forbid). . . . [H]e's rented a place for himself near you, really close to you, and you probably don't know a thing about it, that's why I'm writing to you and I'm sending a special messenger with this letter, so that you'll be informed about it. It's true and you must rent new places outside the town for yourselves and for all our brothers and sons of Israel. . . . We know for sure uncle has got this place almost ready for you. You must know about it, you must find some way out. . . . Uncle is planning to hold this wedding as soon as possible. . . . Go into hiding. . . . Remember—we are holy sacrifices, "and if some is left till morning . . ."

Historian Emanuel Ringelblum (who wrote *Polish-Jewish Relations During the Second World War* while hiding in his Warsaw bunker) and other members of the Underground knew exactly what the letter meant. The cryptic last sentence referred to the Passover instructions in Exodus 12:10: any leftovers of the sacrificial lamb were to be *burnt*. Soon news came from Chełmno of Jews being gassed in vans, and refugees from Wilno told of massacres in other towns as well. Such atrocities still seemed impossible to believe until a man who had escaped the gas chamber and hidden in a freight car all the way to Warsaw told people in the Ghetto what he'd witnessed. Even though the Underground then spread news of Treblinka, some people argued the Nazis wouldn't visit the same bestiality on a city as important as Warsaw.

On July 22, 1942, the liquidation of the Ghetto began on Stawki Street, with 7,000 people herded to the train station, loaded into chlorinated red cattle cars, and delivered to the gas chambers at Majdanek. For this so-called "resettlement in the east," they were allowed to pack three days' worth of food, all their valuables, and thirty-three pounds of personal luggage. Between July and September of 1942, the Nazis shipped 265,000 Jews from Warsaw to Treblinka, leaving only 55,000 behind in the Ghetto, where a Jewish Fighting Organization, known as ZOB (Żydowska Organizacja Bojowa), arose and prepared for combat. To still the doomed as long as possible, the train station at Treblinka posted arrival and departure times for trains, though no prisoners ever left. "With great precision, they started to reach their insane goal," Antonina wrote. "What looked at first like one individual's bloodsucking instinct soon became a well-designed method to destroy whole nations."

Another neighbor of theirs who, like Szymon Tenenbaum and Rabbi Shapira, chose to stay in the Ghetto when offered escape, was pediatrician Henryk Goldszmit (pen name: Janusz

Korczak), who wrote autobiographical novels and books for parents and teachers with such titles as *How to Love a Child* and *The Child's Right to Respect.* To the amazement of his friends, fans, and disciples, Korczak abandoned both his literary and medical careers in 1912 to found a progressive orphanage for boys and girls, ages seven to fourteen, at 92 Krochmalna Street.

In 1940, when Jews were ordered into the Ghetto, the orphanage moved to an abandoned businessmen's club in the "district of the damned," as he described it in a diary written on blue rice-paper pages that he filled with details of daily life in the orphanage, imaginative forays, philosophical contemplations, and soul-searching. It's the reliquary of an impossible predicament, revealing "how a spiritual and moral man struggled to shield innocent children from the atrocities of the adult world during one of history's darkest times." Reportedly shy and awkward with adults, he created an ideal democracy with the orphans, who called him "Pan Doctor."

There, with wit, imagination, and self-deprecating humor, he devoted himself to a "children's republic" complete with its own parliament, newspaper, and court system. Instead of punching one another, children learned to yell "I'll sue you!" And every Saturday morning court cases were judged by five children who weren't being sued that week. All rulings rested on Korczak's "Code of Laws," the first hundred of which parsed forgiveness. He once confided to a friend: "I am a doctor by training, a pedagogue by chance, a writer by passion, and a psychologist by necessity."

At night, lying on his infirmary cot, with remnants of vodka and black bread tucked under his bed, he would escape to his own private planet, Ro, where an imaginary astronomer friend, Zi, had finally succeeded in building a machine to convert radiant sunlight into moral strength. Using it to waft peace throughout the universe, Zi complained that it worked everywhere except on "that restless spark, Planet Earth," and they debated whether Zi should destroy bloody, warmongering Earth, with Doctor Pan pleading for compassion given the planet's youth.

His blue pages stitched together sensations, fancies, and marauding ideas alike, but he didn't relate sinister Ghetto events, for example, the deportations to the death camps that began on July 22, his sixty-fourth birthday. Instead of all the clangor and mayhem on that day, he wrote only of "a marvellous big moon" shining above the destitute in "this unfortunate, insane quarter."

By then, as photographs show, his goatee and mustache had grayed, bags terraced beneath intense dark eyes, and though he often endured "adhesions, aches, ruptures, scars," he refused to escape from the Ghetto, leaving the children behind, despite many offers of help from disciples on the Aryan side. It creased him to hear the starving and suffering children compare their ills "like old people in a sanitarium," he wrote in his diary. They needed ways to transcend pain, and so he encouraged prayers like this one: "Thank you, Merciful Lord, for having arranged to provide flowers with fragrance, glow worms with their glow, and to make the stars in the sky sparkle." By example, he taught them the mental salve of mindful chores, like the slow attentive picking up of bowls, spoons, and plates after a meal:

When I collect the dishes myself, I can see the cracked plates, the bent spoons, the scratches on the bowls. . . . I can see how the careless diners throw about, partly in a quasi-aristocratic and partly in a churlish manner, the spoons, knives, the salt shakers and cups. . . . Sometimes I watch how the extras are distributed and who sits next to whom. And I get some ideas. For if I do something, I never do it thoughtlessly.

Inventing both silly games and the ramparts of deeper play, he decided one day to stage a drama inspired by his affection for Eastern religion, *The Post Office*, by the Indian author Rabindranath Tagore. That production now assumes the potency of symbol, opening as it did on July 18, just three weeks before the children were shipped to Treblinka. In the play, a bedridden boy named Amal suffers in a claustrophobic room and dreams of flying to a land where a king's doctor can cure him. By the play's end the royal doctor appears, heals him, flings open the doors and windows, and Amal beholds a circus of stars. Korczak said he chose the play to help the trapped, terrified children accept death more serenely.

Anticipating their calamity and fright when deportation day came (August 6, 1942), he joined them aboard the train bound for Treblinka, because, he said, he knew his presence would calm them—"You do not leave a sick child in the night, and you do not leave children at a time like this." A photograph taken at the Umschlagplatz (Transshipment Square) shows him marching, hatless, in military boots, hand in hand with several children, while 192 other children and ten staff members follow, four abreast, escorted by German soldiers.

Korczak and the children boarded red boxcars not much larger than chicken coops, usually stuffed with seventy-five vertical adults, though all the children easily fit. In Joshua Perle's eyewitness account in *The Destruction of the Warsaw Ghetto*, he describes the scene: "A miracle occurred, two hundred pure souls, condemned to death, did not weep. Not one of them ran away. None tried to hide. Like stricken swallows they clung to their teacher and mentor, to their father and brother, Janusz Korczak."

In 1971, the Russians named a newly discovered asteroid after him, *2163 Korczak*, but maybe they should have named it *Ro*, the planet he dreamed of. The Poles claim Korczak as a martyr, and the Israelis revere him as one of the Thirty-Six Just Men, whose pure souls make possible the world's salvation. According to Jewish legend, these few, through their good hearts and good deeds, keep the too-wicked world from being destroyed. For their sake alone, all of humanity is spared. The legend tells that they are ordinary people, not flawless or magical, and that most of them remain unrecognized throughout their lives, while they choose to perpetuate goodness, even in the midst of inferno.

Chapter 21

After the great deportations of July 1942, the shape and nature of the Ghetto changed from a congested city of ever-crowded streets into a labor camp full of German workshops policed by the SS. In its large, vastly depopulated southern neighborhood known as "the wild Ghetto," a special corps, the *Werterfassung*, busily salvaged what it could from abandoned belongings and remodeled the deserted homes for German use, while the remaining 35,000 or so Jews were resettled in housing blocks near the shops and escorted to and from work by guards. In reality, another 20,000 to 30,000 "wild" Jews lived in hiding in the Ghetto, staying out of sight, traveling through a maze of subterranean tunnels that led between buildings, and surviving as part of a labyrinthine economy.

Autumn of 1942 also heralded a new Underground group the Żabińskis found immensely helpful: Zegota, cryptonym for the Council to Aid the Jews, a cell founded by Zofia Kossak and Wanda Krahelska-Filipowicz, with the mission of helping Jews hidden in Polish homes. Although its formal name was the Konrad Zegota committee, there was no Konrad Zegota. Zofia Kossak (code name "Weronika"), a noted author and conservative nationalist, mingled freely with the upper classes, especially the landed gentry, and had close friends in the Catholic clergy. In contrast, Krahelska-Filipowicz, editor of the art magazine *Arkady*, was a Socialist activist, wife of a former ambassador to the United States, and well acquainted with military and political leaders of the Underground. Between them, they knew scores of people, and the others they recruited also had a wide network of professional, political, or social contacts. That was the point, to create a human lattice from all corners of society. Aleksander Kamiński, for example, figured in the popular Polish Scouts Association before the war, Henryk Wolinski belonged to the Polish Bar Association, and left-wing Zionist Party member and psychologist Adolf Berman headed Centos, a child welfare organization in the Ghetto. The Writers' Union, the Underground Journalists Association, the Democratic Doctors' Committee, and labor unions comprising railway, tramway, and sanitation department workers all aided Zegota. As Irene Tomaszewski and Tecia Werbowski point out in *Zegota: The Rescue of Jews in Wartime Poland*: "The people of Zegota were not just idealists but activists, and activists are, by nature, people who know people."

Drawing together a consortium of Polish Catholic and political groups, Zegota's sole purpose was rescue, not sabotage or fighting, and, as such, it was the only organization of its kind in occupied Europe during the war, one that historians credit with saving 28,000 Jews in Warsaw. Its headquarters at 24 Zurawia Street, run by Eugenia Wąsowska (a bookbinder and printer) and lawyer Janina Raabe, kept office hours twice a week and also provided temporary shelter for some people on the run. Conspiring with the Polish Underground and

Resistance, it supplied the Żabińskis' villa with money and false documents, and scoured outlying towns for houses where the zoo's Guests could ride out the war. Keeping one person alive often required putting a great many in jeopardy, and it tested them nonstop, as they resisted both propaganda and death threats. Yet 70,000–90,000 people in Warsaw and the suburbs, or about one-twelfth of the city's population, risked their lives to help neighbors escape. Besides the rescuers and Underground helpers, there were maids, postmen, milkmen, and many others who didn't inquire about extra faces or extra mouths to feed.

When Marceli Lemi-Lebkowski, a well-known lawyer and activist, arrived at the zoo with false documents provided by the Underground and "important clandestine missions to fulfill," he and his family pretended to be refugees from the east who wanted to rent two rooms, one for his sick wife, and one for their two daughters, Nunia and Ewa. Marceli would have to live in another safe house and visit them from time to time, because a new man about the villa might be hard to explain—not so a sick woman and her daughters. Their rent bought coke to heat the upstairs bedrooms, which meant more people could lodge in the villa, among them Marek and Dziuś, two young boys serving in the Underground army's Youth Sabotage Group. The boys had left memorial flowers at sites German soldiers frequently used for shooting Poles, and scrawled on walls and fences "Hitler will lose the war! Germany will die!"—deadly offenses.

That winter, some trustworthy legal tenants paid rent, but mainly the villa embraced people lost between worlds and on the run from the Gestapo. In time, the Guests included

Irena Mayzel, Kazio and Ludwinia Kramsztyk, Dr. Ludwig Hirszfeld (a specialist in communicable diseases), Dr. Roza Anzelówna from the National Hygiene Institute, the Lemi-Lebkowski family, Mrs. Poznańska, Dr. Lonia Tenenbaum, Mrs. Weiss (wife of a lawyer), the Keller family, Marysia Aszer, journalist Maria Aszerówna, Rachela Auerbach, the Kenigswein family, Drs. Anzelm and Kinszerbaum, Eugienia "Genia" Sylkes, Magdalena Gross, Maurycy Fraenkel, and Irene Sendler, among many others—according to Jan, about three hundred in all.

As if invisible ink ran through their veins, Jewish and Polish outlaws only appeared indoors, after hours, where Guests and tenants fused into a single family. As a result, Antonina's daily chores increased, but she also had more helpers, and she enjoyed having the two young Lemi-Lebkowska girls around, quickly discovering how little they knew about housework, and schooling them "rigorously" in the wifely trades.

A zoo without animals equaled a waste of land to the Nazis, who decided to build a fur farm on the grounds. Not only would the fur warm German soldiers fighting on the eastern front (they'd already confiscated all fur from the Ghetto Jews for this purpose), extras could be sold to help finance the war. For efficiency they put a Pole in charge of it: Witold Wroblewski, an elderly bachelor used to living alone with fur farm animals. Like the outcast in Mary Shelley's *Frankenstein*, he would enviously watch those inside the warm, comfortable villa, "full of light and the smell of baking bread," he later told Antonina. One day, to Jan and Antonina's surprise and distress, he arrived at their door and, without any niceties or discussion, declared that he was moving in.

Luck favored the Żabińskis, who soon discovered that "Fox Man," as they came to call him, was a Pole raised in Germany who felt sympathy for their mission and could be trusted. By far the most eccentric human in the villa, he arrived with a female cat, Balbina, and what Antonina referred to as "several inseparable parakeets," but nothing else, no personal belongings. That made quick work of moving him into Jan's old study, and he paid with badly needed coke and coal to heat the house. Though it surely impeded his life as a businessman, Fox Man couldn't abide calendars or clocks, street names or numbers; and sometimes he slept on the floor between his desk and his bed, as if fatigue simply overtook him and he hadn't the energy to lurch a step farther. When housemates learned that he had played piano professionally before the war, he entered the Żabińskis' inner circle, because, as Magdalena liked to say: "The House Under a Crazy Star respects artists above all." Though everyone nagged him to play piano, he kept refusing, then one day, at exactly 1 A.M., he emerged from his bedroom, padded quietly to the piano, and suddenly began playing nonstop until morning. After that, Magdalena organized regular piano recitals in the evenings, after curfew, and his Chopin and Rachmaninoff made a wonderful change from the frantic bars of "Go, go, go to Crete!"

Antonina often wrote about Fox Man's gray cat, Balbina, whom she described as appropriately sluttish ("always getting married like a good, normal cat"). But every time Balbina had kittens, Fox Man would snatch them from the basket and replace them with newborn foxes for her to nurse. Antonina doesn't say what became of the kittens, which he may have fed to the fur farm's omnivorous raccoon dogs (ranched for their gray fur with raccoon-like markings). According to breeders, a female fox should only nurse a few pups at one time, to ensure all grow thick, healthy coats; using Balbina as a wet nurse for the extra pups struck him as an ideal if somewhat impish solution. "The first day was always the hardest for her," Antonina noted, "she could swear that she gave birth to kittens, but on the second day she knew it was only her imagination."

Understandably confused by their odd scent and snarls, the cat discovered the baby foxes had ravenous appetites, and after lots of licking and feeding they finally began to smell like her, though her repeated attempts to school them in the feline arts mainly failed. Meowing around them in "a quite distinguished tone of voice . . . to teach them how normal cats should speak," she never did persuade them to meow back, and their loud barking constantly startled her. "In her cat's heart, she was ashamed that they barked," Antonina mused, adding that the offspring were loudmouths with "high tempers." But they did master an agile cat-leap onto tables, cabinets, and tall bookcases, and villa-ites often found a baby fox, curled up like a Bavarian soup tureen, napping atop the piano or a chest of drawers.

Favoring live food, Balbina hunted outside each day to feed her brood, diligently dragging home birds, rabbits, meadow mice, and rats, though, as she soon discovered, she needed to hunt nonstop to still their lidless hunger. Outside, she led the way—a small, thin tabby followed by offspring three times her size with long snouts and fluffy black tails ending in white flowers. She taught them how to stalk prey while crouching low like a sphinx, how to pounce on game, and if one strayed, she meowed harshly until the young fox dutifully trotted back to the fold. Whenever the fox pups spied a chicken, they stalked it, crawling fast on their bellies, then pouncing with sharp teeth to rip it apart, snarling as they fed, while Balbina kept her distance and watched.

After "giving birth" to several broods of baby foxes, tiring and confusing as that was, Balbina finally got used to their alien ways, and they became half-cat, she half-vixen. Praising the cat's good-citizenry of never attacking housemates, Antonina wrote: "It's as if she has her own moral code." She spared Fox Man's parakeets, even when he released them from their cage; Wicek the rabbit didn't tempt her, nor did Kuba the chick; she didn't bother hunting the invading mouse or two; and if a stray bird flew into the house (a bad omen), she'd eye it lazily. But one new arrival did rekindle Balbina's feral instincts.

In the spring, a neighbor brought a strange orphan for Ryś's royal zoo—a paunchy baby muskrat with glossy brown fur, yellow-beige belly, long scaly tail, and tiny black eyes. Webbed front paws with fingers help muskrats build lodges, hold food, or dig burrows; when they swim, fringed hind feet make strong canoe-paddle sweeps. Oddest of all, perhaps, four sharp chisel-like front teeth protrude beyond the cheeks and lips, so that a muskrat can eat stems and roots, bulrushes and cattails while underwater, without opening its mouth.

Antonina found the creature fascinating and gave it a large cage on the porch, adding a glass developing tray from an old darkroom as wading pool, since muskrats are native-born swimmers. Ryś named it Szczurcio (Little Rat), and soon it learned its name and adapted to life in the three-ring villa, spending its days sleeping, eating, or wallowing. Wild muskrats don't tame easily, but in a few weeks Szczurcio let Ryś open the cage, carry him around, and pet or scratch his fur. While Szczurcio slept, Balbina would circle the cage like a mountain lion, searching for a way in. Awake, he tormented her by playing in the little tub incessantly and splashing her with water, which she hated. No one knew why the muskrat tempted Balbina so, but whoever fed Szczurcio or cleaned his cage had to lock the door afterward with small twists of wire.

Antonina enjoyed watching the muskrat's "exquisite toilette"—each morning, Szczurcio would dunk his face in the pan of water and snort heartily, blow air out his nose, then splash his face with wet paws like a man preparing to shave, and wash for a long time. After that he would climb into the tub and stretch out on his belly, turn onto his back, and roll over several times. Finally he left the bathtub and shook his fur like a dog, splashing mightily. Strangely enough, he often climbed the wall of the cage and sat on the perch like the cage's previous occupant, Koko the cockatoo. There, using his fingers, he would carefully comb water through his fur. Visitors found it a little odd to see a muskrat perching and preening like a bird, but the villa held a bizarre crew even at the quietest times, and he was Ryś's new favorite pet. After his morning ablutions, Szczurcio would eat a carrot, potato, dandelions, bread, or grain, though he no doubt craved the branches, bark, and marsh weeds on which wild muskrats thrive.

When he outgrew the little tub of water, Antonina replaced it with a giant jar Jan had once used in a cockroach study. Szczurcio leapt into the jar when it arrived and splashed with such abandon that Antonina moved his cage into the kitchen, where the floor was ceramic tile and fresh water lay closer to hand.

"You know, Mother," Ryś said one day, "Szczurcio is learning how to open his cage. He's not stupid!"

"I don't think he's quite *that* smart," Antonina replied.

Szczurcio spent hours fiddling with the wire, grabbing the ends with his fingers and trying to untwist them, and after a night's crafty work, he finally succeeded in unknotting the wire and lifting up the sliding door, scrambling down a chair leg to the floor, shimmying up the water pipe, and sliding into the marshlike kitchen sink. Then he leapt atop the stove, climbed onto a warm radiator, and fell asleep. That's where Ryś found him in the morning. Returning him to his cage, Ryś closed the door and knotted the wire even tighter.

Early the next day, Ryś ran through the house to Antonina's bedroom, where he cried out in alarm: "Mom! Mom! Where's Szczurcio? His cage is empty! I can't find him anywhere! Maybe Balbina ate him? I have to go to school, and Dad is at work! Help!"

Still bedridden, Antonina couldn't help much with this dawn crisis, but she deputized Fox Man and the housekeeper, Pietrasia, to launch a search party, and they dutifully scouted all the closets, sofas, easy chairs, corners, boots—any bolt-hole where a muskrat might hide—with no success.

Because she couldn't believe the muskrat had simply "evaporated like camphor," she suspected Balbina or Żarka of mayhem, and had cat and dog brought to her bed for close inspection. There she carefully felt their stomachs for any suspicious bulges. If they'd eaten such a large animal—almost the size of a rabbit—surely their bellies would still be swollen. No, they felt slender as usual, so she declared the detainees innocent and released them.

Suddenly Pietrasia ran into her bedroom. "Come quick!" she shouted. "To the kitchen. Szczurcio is in the stove chimney! I started a fire the way I do every morning, and I heard a terrible noise!"

Using her cane, Antonina slowly rose from bed onto her swollen legs, carefully descended the stairs, and hobbled into the kitchen.

"*Szczurcio, Szczurcio,*" she called sweetly.

A scuffling noise in the wall. When a soot-covered head poked from the chimney, she grabbed the fugitive's back and pulled him out, whiskers coated in grime, front paws singed. Gently, she washed him in warm water and soap, over and over, trying to remove the cooking grease from his fur. Then she applied ointment to his burns and placed him back in his cage.

Laughing, she explained that a muskrat builds a lodge by heaping plants and mud into a mound, then digging a burrow from below water level. This muskrat wanted a lodge not a cage, she said, and who could blame it for creating a facsimile world? He had even bent the metal burners to fashion an easier route into the chimney.

When Ryś returned from school that afternoon, he was thrilled to find Szczurcio back in his cage, and at dinner, as people carried food to the table, Ryś regaled all with the adventures of Szczurcio and the stovepipe. One little girl laughed so hard that she tripped on her way from the kitchen, spilling a full bowl of hot soup over Fox Man's head and onto Balbina, who had been sitting in his lap. Springing from his chair, Fox Man bolted into his bedroom, followed by his cat, and closed the door. Ryś ran after him, spied through the keyhole, and whispered regular reports:

"He took off his jacket!"

"He's drying it using a towel!"

"Now he's drying Balbina!"

"He's drying his face!"

"Ooooh! No! He opened the cage with his parakeets!"

At this point, Magdalena couldn't stand the suspense any longer and flung open the door. There stood Fox Man, the house concert master, column-like in the middle of the room, with parakeets circling his forehead like merry-go-round animals. After a few moments they landed on his head and started digging through his hair, pulling out and eating the soup noodles. At last Fox Man noticed the crowd in the doorway, silent and agog, waiting for some explanation.

"It would be a pity to waste such good food," he said of the bizarre scene, as if he'd found the only and obvious thing to do.

Chapter 22
Winter, *1942*

Time usually glides with an incoherent purr, but in the villa it always quickened as curfew hour approached, when a kind of solstice took place and the sun stopped on the horizon of Antonina's day, with minutes moving as slow as mummers: one, a stretched pause, then another. Because anyone who didn't make it home by curfew risked being arrested, beaten, or killed, the hour acquired a pagan majesty. Everyone knew curfew horror stories like that of Magdalena's friend, painter and prose writer Bruno Schulz, gunned down by a spiteful Gestapo officer in Drohobycz, on November 19, 1942. Another Gestapo officer, Felix Landau, who admired Schulz's macabre, sometimes sadomasochistic paintings, had given him a pass out of the Ghetto to paint frescoes of fairy tales on his son's bedroom walls. One day, Landau killed a Jewish dentist under the protection of Günther, another officer, and when Günther spotted Schulz in the Aryan quarter after curfew, walking home with a loaf of bread under his arm, he shot him in retaliation.

If everyone arrived safely, Antonina celebrated another day without mishap, another night unmauled by monsters in the city's labyrinths. Curfew twilight tormented Ryś, so she allowed him to stay up and await the homecomings; then he could fall asleep peacefully, his world intact. Years of war and curfews didn't alter that; he still anxiously awaited his father's return, indispensable as the moon's. Respecting this, Jan would go straight to Ryś's room, remove his backpack, and sit for a few minutes to talk about the day, often producing a little treasure tucked in a pocket. One night his backpack bulged as if it had iron ribs.

"What do you have there, Papa?" Ryś asked.

"A tiger," Jan said in mock fear.

"Don't joke, what's really in there?"

"I told you—a *dangerous* animal," his father said solemnly.

Antonina and Ryś watched Jan remove a metal cage containing something furry, shaped like a dwarf guinea pig, mainly chestnut in color, with white cheeks and spots on its sides like a Sioux horse.

"If you'd like to have him, he's yours!" Jan said. "He's a son of the hamster couple I have at the Hygiene Institute. . . . But if I give him to you, you're not going to feed him to *Balbina,* are you?" Jan teased.

"Papa, why do you talk to me like I'm a little child?" Ryś said, offended. He'd had all sorts of pets in the past, he argued, and hadn't done anything wicked with them.

"I'm very sorry," Jan said. "Take good care of him, keep a close eye on him, because he's the only survivor from a litter of seven. Unfortunately, the others were killed by their mother before I could stop her."

"What a horrible mother! Why do you keep her?"

"All hamsters have this cruel instinct, not only his mother," Jan explained. "A husband can kill his wife. Mothers chase their youngsters away from the burrow and don't care for them anymore. I didn't want to deprive the babies of their mother's milk too early, but unfortunately I miscalculated the best moment and was only able to save this one. I don't have time to take care of him at the lab, but I know you will do a great job."

Antonina wrote that she and Jan found it hard to decide how much to tell a small child about the amoral, merciless side of nature, without scaring him (the war offered frights enough), but they also felt it important that he know the real world and learn the native ways of animals, explicably vicious or inexplicably kind.

"I've read so many stories about hamsters," he said, disappointed, "and I was so sure they were nice, hardworking animals who collected grain for the winter. . . ."

"Yes, that's true," Jan said reassuringly. "During the winter he hibernates, just like a badger, but if he happens to wake up hungry during winter, he can eat the grain and go back to sleep until spring."

"It's winter now, so why is this hamster awake?"

"Animals behave differently in the wild. We make captive ones live on a schedule that's unnatural to them because it's easier for us to take care of them, and that disturbs their normal sleep rhythms. But even though this hamster is awake, his pulse and breathing are a lot slower than they will be in the summer. You can check this for yourself—if you cover his cage, he'll fall asleep almost immediately."

Ryś drew a blanket over the cage and the hamster crept into a corner, settled back on its haunches, tucked its head down on its chest, covered its face with its front paws, and fell into a deep sleep. In time, Antonina judged him a "quite self-centered" little being, and "a noisy glutton" who "preferred his own company and an easy life." In a household that porous, where animal time and human time swirled together, it made sense to identify the passage of months not by season or year but by the stay of an influential visitor, two- or four-legged. To Antonina, the hamster's arrival "started a new era on our Noah's Ark, which we later called the 'Hamster Era.'"

Chapter 23

New year, 1943, approached with Antonina still mainly bedridden, and after three months, cabin fever and lack of exercise had depleted her body and spirits. She usually kept the door to her bedroom open so that she could join, however remotely, the stir of the house, its mingling smells and sounds. On January 9, when Heinrich Himmler visited Warsaw, he condemned another 8,000 Jews to "resettlement," but by now everyone understood that "resettlement" meant death, and instead of lining up as ordered, many hid while others ambushed the soldiers and dashed across the rooftops, creating just enough friction to curb deportations for several months. Surprisingly, sketchy telephone service continued, even to some bunkers, though it's hard to imagine why the Germans allowed it, unless they figured clever electricians could hook up illegal phones anyway or the Underground had its own telephone workers.

Before dawn one day, the Żabińskis awoke, not to a chorus of gibbons and macaws as they used to, but to a jangling telephone and a voice that seemed to come from the far side of the moon. Maurycy Fraenkel, a lawyer friend who lived in the dying Ghetto, asked if he could "visit" them.

Although they hadn't heard from him in quite a while, on at least one occasion Jan had visited him in the Ghetto, and they knew him as Magdalena's "dearest friend," so they quickly agreed. Antonina noted that several nerve-trampling hours followed for Magdalena,

> whose lips were blue, and her face so white that we could see many freckles, normally almost invisible. Her strong, ever-busy hands were trembling. The sparkle had vanished from her eyes, and we could read only one painful thought on her face: "Will he be able to escape and come here?"

He did escape but arrived a gnarled specimen, bent over like a gargoyle from the Other Side, as people sometimes referred to the Ghetto, a Yiddish term, *sitre akhre*, for the dim world where demons dwell and zombies wear "a husk or shell that has grown up around a spark of holiness, masking its light."

The unbearable weight of ghetto life had physically crippled him—his head hung low between curved shoulders, his chin rested on his chest, and he breathed heavily. Swollen red from frost, his nose glowed against a pale, sickly face. When he entered his new bedroom, in a dreamy sort of way he dragged an armchair from beside the wardrobe to the darkest corner of the room, where he sat hunched over, shrinking himself even more, as if he were trying to become invisible.

"Will you agree to have me here?" he asked softly. "You will be in danger. . . . It is so quiet here. I can't understand. . . ." That was all he could manage before his voice trailed away.

Antonina wondered if his nervous system, adapted to the hurly-burly of Ghetto life, found this sudden plunge into calm and quiet unnerving, if it sapped more energy from him than the distressed world of the Ghetto had.

Born in Lwów, Maurycy Paweł Fraenkel had a passion for classical music, many composers and conductors as friends, and he had often organized small, private concerts. As a young man he studied law and moved to Warsaw, where he met Magdalena Gross, whose gift he greatly admired, at first becoming her patron, then close friend, and finally sweetheart. Before the war, she had brought him to the zoo, which he relished, and he had helped the Żabińskis buy several boxcars full of cement to use in zoo renovations.

Maurycy soon grew used to life across the river from the lurid Ghetto, and as he ventured out of corners and shadows, Antonina wrote that his backbone seemed to straighten a little, though never completely. He had a sarcastic sense of humor, though he never laughed out loud, and a huge smile would light his face until his eyes scrunched and blinked behind thick glasses. Antonina found him

calm, kind, agreeable, and gentle. He didn't know how to be aggressive, frightful, or disagreeable even for a second. This was why he moved to the Ghetto when told to, without thinking twice about it. After he experienced the full tragedy of being there, he tried to commit suicide. By luck, the poison he used was too stale to work. After that, with nothing to lose, he decided to risk an escape.

Without documents he couldn't register anywhere, so officially he ceased to exist for a long time, living among friends but gaunt and ghostly, one of the disappeared. He had lost many voices: the lawyer's, the impresario's, the lover's, and it isn't surprising that he found speaking or even coherence difficult.

While Antonina lay ill, Maurycy sat next to her bed for hours, slowly recovering his spiritual balance, Antonina thought, as well as the energy to talk again. What weighed heaviest was the colossal risk he created just by being there, and he often referred to Governor Frank's threat of October 15, 1941, the decree that all Poles hiding Jews would be killed. Every Jew receiving help had to deal with this painful issue, including the dozen hidden in the villa and the rest in the animal houses, but Maurycy was especially bothered by the burden he added to the Żabińskis' lives. It was one thing to expose *himself* to danger, he told Antonina, but the thought of spreading an epidemic of fear throughout the zoo, the hub of so many lives, piled on more guilt than he could shoulder.

In Antonina's bedroom, shelves and drawers recessed into white walls, and the bed nestled in a shallow alcove, from which it jutted like a well-upholstered pier. All the furniture had been crafted from silver birch, a plentiful tree in Poland, both hard and durable, a pale wood whose fibers vary from plain to flamelike, with here and there brown knots and fine brown traces of insects that once attacked the cambium of the living tree.

On the south side of the room, beside tall windows, a glass door opened onto the wraparound terrace; and on the north side, three white doors led to the hallway, the attic, and the step-in closet where Guests hid. Instead of the lever handles of the villa's other doors, the closet bore a high keyhole, and though it offered little space inside, a Guest could curl up there among the glide of fabrics and Antonina's comforting scent. Because the closet opened on both sides like a magician's trunk, bunched clothes concealed the opposite door whichever way one looked. As safety hatches go, it served well, especially since its hallway door began a foot or so above the floor, suggesting only a shallow cupboard, which a pile of laundry or a small table could easily disguise.

One day Maurycy, seated in a bedside chair, heard the housekeeper Pietrasia on the stairs and he hid in the closet, nestled among Antonina's polka-dot dresses. As Pietrasia left the room, Maurycy quietly emerged and sat down, but before Antonina could say a word Pietrasia opened the door and rushed back in with a housekeeping question she'd forgotten to ask. Seeing a stranger, she stopped abruptly, breathed hard, and frantically crossed herself.

"So you will continue to take salicylic acid," Maurycy said to Antonina in a doctorly tone, and delicately holding her wrist, he added: "And now I will check your pulse." Later, Antonina wrote that her anxious pulse wasn't hard to feel, and that his own had pounded down to his fingertips.

Pietrasia studied their faces, finding them calm, and shook her head in confusion. Mumbling that she must have had some sort of vision problem or blackout, she left the room, rubbing her brow and shaking her head as she went downstairs.

Antonina called Ryś and said: "Please bring me the *doctor's* coat and hat and let him out of the house by the kitchen door, so that Pietrasia will see him leaving. After that, call her to check on the chickens. Do you understand?"

Ryś blinked his eyes, thought awhile, and then a smile crept over his face. "I'll tell her that this morning I accidentally let a chicken out and we have to find it. Then the *doctor* can sneak back in through the garden door. That would work."

"Thank you for being so smart," Antonina told him. "Now hurry!"

From then on, Maurycy only roamed the house at night, after the housekeeper had left for the day and he could safely prowl downstairs, as if on forbidden tundra. Every evening, Antonina found him walking back and forth across the living room, slowly, reverently, so that he "would not forget how to walk," he explained. At some point, he'd pause to check on the hamster he'd befriended, before joining other Guests for Fox Man's piano concert.

One evening, between Rachmaninoff preludes, Fox Man took Maurycy aside and said, "Doctor, I'm bad at paperwork, and some of it's in German—a language I don't speak well at all. My fur business is growing and I really need a secretary. . . . Maybe you could help me?"

Maurycy had once confided to Antonina that, in seclusion, using an unfamiliar name, he felt like a phantom. This offer of Fox Man's meant Maurycy could become real again, with papers and mobility, and, best of all, residency status in the villa as an employee of the fur farm. Becoming real was no small accomplishment, since occupation ushered in an overgrowth and undergrowth of official identity cards and documents—bogus working papers, birth certificate, passport, registration card, coupons, and passes. His new papers declared him to be Paweł Zieliński, the official secretary of the fox farm, and so he rejoined the household as a lodger, which also meant he didn't have to hide in the upstairs closet, a space now available for another Guest. Becoming real brought psychological changes, too. He slept on a couch downstairs in the hamster's narrow room, adjacent to the dining room, among the rustlings of his favorite pet, and Antonina noticed that his entire mood began to change.

Maurycy told Antonina that every night he prepared his bed slowly with a happiness unknown to him since before occupation, taking pleasure in the simple acts of carefully folding his only suit, frayed as it was, and laying it over a chair beside his own bookshelf, occupied by the handful of books he had salvaged from his old life, in a house where he could sleep unmolested, surrounded by a surrogate family whose presence padded his existence.

For a great many people, the Ghetto had erased the subtle mysticism of everyday life, such reassuring subliminals as privacy, agency, and above all the faith that allows one to lie down at night and surrender easily to sleep. Among the innocence of hamsters, Maurycy slept near his books, with documents that bestowed the status of being real, and, best of all, under the same roof as his beloved Magdalena. Finding love undemolished, with enough space to exist and his heart still limber, gave him hope, Antonina thought, and even renegade "moments of pleasure and joy, feelings he'd lost in Ghetto life."

On February 2, 1943, the German Sixth Army surrendered at Stalingrad in the first big defeat of the Wehrmacht, but only three weeks later Jews working in Berlin armaments factories were freighted off to Auschwitz, and by mid-March the Kraków Ghetto was liquidated. Meanwhile, the Underground continued attacks of various kinds, 514 since January 1; and on January 18 the first armed resistance began in the Warsaw Ghetto.

During this time of seismic upheaval, more and more Ghetto dwellers washed up on the deck of the villa, arriving weather-beaten, "like shipwrecked souls," Antonina wrote in her diary. "We felt that our house wasn't a light, flimsy boat dancing on high waves, but a Captain Nemo's submarine gliding through deep ocean on its journey to a safe port." Meanwhile, the war storm blew violently, scaring all, and "casting a shadow on the lives of our Guests, who fled from the entrance of crematoriums and the thresholds of gas chambers," needing more than refuge. "They desperately needed hope that a safe haven even existed, that the war's horrors would one day end," while they drifted along in the strange villa even its owners referred to as an ark.

Keeping the body alive at the expense of spirit wasn't Antonina's way. Jan believed in tactics and subterfuge, and Antonina in living as joyously as possible, given the circumstances, while staying vigilant. So, on the one hand, Jan and Antonina each kept a cyanide pill with them at all times, but on the other, they encouraged humor, music, and conviviality. To the extent possible, theirs was a bearable, at times even festive, Underground existence. Surely, in response to the inevitable frustrations brought about by living in close quarters, the Guests uttered Yiddish's famous curses, which run the gamut from graphic ("May you piss green worms!" or "A barracks should collapse on you!") to ornate:

You should own a thousand houses
with a thousand rooms in each house
and a thousand beds in every room.
And you should sleep each night
in a different bed, in a different room,
in a different house, and get up every morning
and go down a different staircase
and get into a different car,
driven by a different chauffeur,
who should drive you to a different doctor
—and he *shouldn't know what's wrong with you either!*

Nonetheless, "I have to admit that the atmosphere in our house was quite pleasant," Antonina confessed in her diary, "sometimes even almost happy." This contrasted sharply with the texture of life and the mood inside even the best hideouts around town. For example, Antonina and Jan knew Adolf Berman well, and most likely read the letter Adolf received in November 1943, from Judit Ringelblum (Emanuel's wife), which told of the mood in a bunker nicknamed "Krysia":

Here a terrible depression reigns—an indefinite prison term. Awful hopelessness. Perhaps you can cheer us up with general news and maybe we could arrange for the last of our nearest to be with us.

Sharing a room, the hamster and Maurycy seemed to find amusement in each other, and Antonina noted how quickly the two became companions. "You know what," Maurycy said one day, "I like this little animal so much, and since my new name is Paweł [Paul], I think his should be Piotr [Peter]. Then we can be two disciples!"

After supper each evening Maurycy turned Piotr loose on the table's polished mesa, where the hamster skittered from plate to plate, whisking up crumbs until his fat cheeks dragged. Then Maurycy would gather him up in one hand and carry him back to his cage. In time, Piotr trusted him enough to float around the house on the carpet of Maurycy's open palm, the pair became inseparable, and villa-mates began referring to Paweł and Piotr collectively as "the Hamsters."

Chapter 24

In the spring of 1943, Heinrich Himmler wished to give Hitler an incomparable birthday present, one to elevate him above all others in Hitler's favor. Himmler, who often held intimate conversations with Hitler's photograph and strove to be Hitler's best and most faithful servant, would have lassoed and gift-wrapped the moon if he could. "For him, I would do anything," he once told a friend. "Believe me, if Hitler were to say I should shoot my mother, I would do it and be proud of his confidence." As a gift, he swore to liquidate the remaining Jews in the Warsaw Ghetto, on April 19, the first day of Passover, an important Jewish holy day, and also the eve of Hitler's birthday.

At 4 A.M., small German patrols and assault squads cautiously entered the Ghetto and caught a few Jews on their way to work, but the Jews somehow managed to escape and the Germans withdrew. At 7 A.M., Major General Jürgen Stroop, commander of an SS brigade, returned with 36 officers and 2,054 soldiers, and roared straight to the center of the Ghetto with tanks and machine guns. To his surprise, he found barricades manned by Jews who returned fire with pistols, several rifles, one machine gun, and many "Molotov cocktails," gasoline-filled bottles bunged up with burning rags. Finns had recently borrowed the idea of the bottle grenade from Franco nationalists, who improvised it during the 1936–39 Spanish Civil War, a time when predinner cocktails slid into vogue among the swanky set. When Russia invaded Finland, the Finns sarcastically named the bomb after Foreign Minister Vyacheslav Mikhailovich Molotov. Though vastly outnumbered and ill-equipped, Jews managed to hold Nazis at bay until nightfall, and again the following day when soldiers reappeared with flamethrowers, police dogs, and poison gas. From then on, about 1,500 guerrillas fought back at every chance.

What Himmler planned as a gift-wrapped massacre became a siege lasting nearly a month, until at last the Germans decided to torch everything—buildings, bunkers, sewers, and all the people in them. Many died in the fires, some surrendered, others committed suicide, and a few escaped to tell and write of the armageddon. Underground newspapers called upon Christian Poles to help escaping Jews find shelter, and the Żabińskis eagerly obliged.

"Nearby, on the other side of the wall, life flowed on as usual, as yesterday, as always," one survivor wrote. "People, citizens of the capital, enjoyed themselves. They saw the smoke from the fires by day and the flames by night. A carousel went round and round beside the ghetto, children danced in a circle. It was charming. They were happy. Country girls visiting the capital rode on the roundabout, looking over the flames of the ghetto," laughing, catching leaves of ash that floated their way, as a loud carnival tune played.

Finally, on May 16, Major General Stroop sent Hitler a

proud report: "The Warsaw Ghetto is no more." According to the *Underground Economic Bulletin* of May 16, 1943, 100,000 apartments burned down, 2,000 places of industry, 3,000 shops, and a score of factories. In the end, the Germans captured only 9 rifles, 59 pistols, and several hundred home-made bombs of various sorts. Seven thousand Jews had been shot outright, 22,000 were shipped to the death mills of Treblinka or Majdanek, and thousands more went to labor camps. To achieve this cost the Germans only 16 dead and 85 wounded.

As everyone at the villa followed news of the Ghetto Uprising, Antonina recorded their mood as "electrified, stunned, helpless, proud." At first, they'd heard that Polish and Jewish flags were hoisted above the Ghetto, then, as smoke and sounds of artillery fire rose, they learned from their friend Stefan Korboński, a high-ranking member of the Underground, that the Jewish Fighting Organization and the Jewish Fighting Union—only 700 men and women—were battling heroically, but "the Germans have removed, murdered, or burned alive tens of thousands of Jews. Out of the three million Polish Jews no more than 10 per cent remain."

Then, one terrible day, a gray rainfall settled on the zoo, a long, slow rain of ash carried on a westerly wind from the burning Jewish Quarter just across the river. Everyone at the villa had friends trapped in that final stage of annihilating Warsaw's 450,000 Jews.

On December 10, just before curfew, after Jan had made it home safely again and Pietrasia had left for the day, Antonina summoned the family, Fox Man, Magdalena, Maurycy, Wanda, and others to the dinner table for their evening soup of borscht, a glossy red beet soup that reflects candlelight and pools like claret on a large silver spoon. Despite the swirling cold that appeared as snow-djinns under the streetlamps, the villa had enough coal to keep everyone warm that winter. In the kitchen, after dinner, while Ryś was changing the water in Szczurcio's bathtub, he heard a quiet knocking. Carefully, he opened the door, then he ran excitedly into the dining room to tell his parents the news.

"Mom," he said, "Sable's daughter and her family are here!"

Mystified, Fox Man set down his newspaper. The fur farm didn't raise sable, a small minklike animal.

"This house is totally crazy!" he said. "You use animal names for people and people's names for animals! I never know whether it's people or animals you're talking about. Who or what is this 'Sable'? I don't know if it's a first name or a code name or a person's name or an animal's name. It's all too confusing!" Then he stood up dramatically and went to his room.

Antonina hurried to the kitchen to greet the new sables in the house: Regina Kenigswein, her husband Samuel, and their two boys—five-year-old Miecio and three-year-old Stefcio. Their youngest, Staś, less than a year old, went to a foundling home run by Father Boduen, because they worried the baby's crying might draw attention. Regina was also "carrying a baby under her heart," as the saying went—she was pregnant with her fourth.

In the summer of 1942, during the mass deportations to concentration camps, with the passages from the courthouse sealed and escape routes through the mazy sewers not yet mapped, Samuel had asked a Catholic friend, Zygmunt Piętak, for help escaping with his family and finding refuge on the Aryan side. A complex web of friends, acquaintances, and chance primed most escapes from the Ghetto, and that held true for the Kenigsweins. Samuel and his friend Szapse Rotholc had joined the Ghetto police force and quickly befriended sympathetic or greedy German guards and Polish smugglers. At night, carrying the sedated children in sacks, the Kenigsweins bribed guards and climbed over the Ghetto wall. At first they were placed in an apartment Piętak had rented for them, where they hid until late 1943. During that entire time, Piętak served as their only contact with the outside world, visiting them often with food and necessities. But when money ran out and they were evicted, Piętak asked Jan if he could lodge the family while the Underground found them shelter elsewhere.

Antonina knew Regina, the daughter of a Mr. Sable (in Polish, Sobol) who had supplied fruit for the zoo animals before the war, a kind, stoop-shouldered man who always wore the same old faded vest, and lumbered beneath heavy baskets of fruits and vegetables. Despite the load, he usually found room in his pockets for extra treats and gifts, like sweet cherries for the monkeys or a yellow apple for Ryś. But the real bridge between the Sable family and the Żabińskis was through Mr. Sable's son, who belonged to the Ghetto labor force and sometimes stole away from his work site and ran to the zoo, where the Żabińskis gave him potatoes and other vegetables to smuggle home. One day, he explained that he'd been reassigned to another work gang *inside* the Ghetto, and implored Antonina to cajole his

German boss into letting him continue working outside. Antonina did, and noted afterward:

"Maybe this *Arbeitsführer* was a good man, or maybe he was just shocked when I told him that without the food Sable took back to the Ghetto his family would starve to death. Using quite good Polish, he said that I should be 'more careful.' But young Sable was allowed to keep working outside the Ghetto and bringing food home to his family for over a month."

Not only had the Żabińskis known Regina as a girl, they had attended her wedding and Jan had worked with her husband, Samuel—to build bunkers. A famous boxer, Samuel Kenigswein used to fight at the Maccabee and Stars sports clubs in Warsaw, and he was also a trained carpenter who helped Zegota create and remodel hideouts. During the war, architect Emilia Hizowa, a central figure in Zegota, invented false walls that slid open at the push of a button, and workmen installed them in flats around the city, where residents took care not to block them with furniture. The ploy worked: The uncluttered passed as the honest and drew no attention.

When the Kenigsweins first arrived at the zoo, their plight stirred Antonina deeply: "I looked at them with tears in my eyes. Poor chicks with big eyes full of fear and sadness looked back at me." Regina's eyes, especially, disturbed her, because they were "the leaden eyes of a young mother doomed to death."

Antonina wrote that she felt a wrenching inside, a tug-of-war between compassion and self-interest, and a kind of embarrassment that she could do so little for them without endangering herself and her own family. Meanwhile, where would the Kenigsweins sleep? For several days, they stayed in the Lions House, then Regina and the children moved through the Pheasant House tunnel into the villa. Antonina found a large warm sheepskin coat and a pair of boots for Samuel, and before nightfall, he stole into the wooden Pheasant House and they locked him inside. The next morning, before the housekeeper arrived, Regina and her children quietly moved upstairs to a bedroom on the second floor, where they would stay for two months. When Antonina praised the children for making so little fuss or noise, she learned that a secret Ghetto school had taught them games to play in small areas, the quietest ways to move, and how to lie down fluently in as few bends as possible.

The fox farm employed many strangers; unknown boys sometimes stopped by the kitchen, looking for handouts; policemen often visited, too. What's more, the housekeeper couldn't really be trusted, nor could the Żabińskis tell her why their appetites suddenly swelled. Because they couldn't steal food from the kitchen without her noticing, they went to her looking ravenous, empty plate in hand, asking for seconds, thirds, fourths. As a servant, it wasn't her place to comment on their robust change in eating habits, but now and then Antonina heard her muttering: "I can't believe how much they eat! I've never seen anything like it!" When she wasn't looking, Ryś sneaked plates and bowls upstairs and downstairs, one after the other. Sometimes Jan or Antonina would tell him: "The lions need to be fed," or the "pheasants," "peacocks," and so on, and Ryś would carry food to the caged Guests. But to play it safe, Antonina fired the housekeeper, replacing her with a woman named Franciszka, the sister-in-law of an old friend of Jan's, someone they trusted, though even she never knew all the planes of existence and resistance in the three-dimensional chess game of villa life.

Chapter 25

1943

In the middle of December, Jan secured fresh lodgings for the Kenigsweins with engineer and former career officer Feliks Cywiński, who had fought beside Jan during World War I and now worked closely with him in the Underground. Married with two children, Cywiński hid many people in his apartments at 19 and 21 Sapieżyńska Street, at his sister's flat, his parents', and in the upholstery shop of a friend (who closed it for a while, supposedly for renovations). There, he fed as many as seventeen people, providing separate pots and dishes for those who kept kosher, and bringing in medicine and an Underground doctor when necessary. A secret "Coordinating Committee of Democratic and Socialist Doctors," set up in 1940, included over fifty physicians who cared for the sick or wounded, and they also published their own monthly periodical, in which they debunked Nazi propaganda about racial purity and disease. Once a month, Cywiński would move the Jews hiding with him to the zoo or some other safe house, so that he could invite neighbors and friends to his home, proving he'd nothing to hide. When his money ran out, Feliks went into debt, sold his own home, and used the profits to rent and furnish four more apartments for hiding Jews. Like the Kenigsweins, his charges often arrived from the zoo and stayed only a day or two, while documents were procured and other homes found.

Moving the Kenigsweins created a new problem for Antonina and Jan—how to transfer so many people without attracting notice. Antonina decided to lessen the risk by bleaching their black hair blond, since many Germans and Poles, too, assumed all blonds came from Scandinavian stock and all Jews had dark hair. This fallacy endured, even when jokes circulated about Hitler's non-Aryan mustache and dark hair. From photographs and a comment of Jan's, one learns that, at some point, Antonina had bleached her own brown hair, but that only meant lightening it several shades, not transforming it from shadow-black to citrine, and so she consulted a barber friend who gave her bottles of pure peroxide and a recipe. She needed a recipe because, as Emanuel Ringelblum emphasized: "In practice, it turned out that platinum blondes gave rise to more suspicion than brunettes."

One day, she led the Kenigsweins into the upstairs bathroom, locked the door, and stationed Ryś outside as guard. Using cotton balls soaked in diluted peroxide, she rubbed down one head after another, creating scalded red scalps and blistered fingers, but still their hair wouldn't yield a blond, even if she strengthened the caustic solution. When she opened the door at last, her victims emerged with brassy red hair.

"Mom, what did you do?" Ryś asked in alarm. "They all look like squirrels!" From that day on, "Squirrels" became the Kenigsweins' code name.

At night, Jan escorted the Kenigsweins through the basement tunnel to the Pheasant House and downtown to Feliks's home on Sapieżyńska Street. There, in times of danger, refugees would climb into a bunker whose entrance was a camouflaged opening in the bathroom, tucked in a recess behind the bathtub. Feliks didn't know that Regina was pregnant until she went into labor one day, and then, since it was already after curfew, too late to call a doctor, the midwifery fell to him. "My happiest moment," he said in a postwar interview, "was when a child was born literally into my hands. This was during the final destruction of the Warsaw Ghetto. The atmosphere in the town was very tense and the terror was raging at its ugliest, as German gendarmes and blackmailers penetrated the terrain and searched it thoroughly looking for escaping Jews." Feliks cared for them until the Warsaw Uprising in 1944, when Samuel Kenigswein, a World World I veteran, spearheaded a battalion of his own.

Elsewhere in the city, other rescuers were also resorting to cosmetic tricks to disguise Jews, with some salons specializing in more elaborate ruses. For instance, Dr. Mada Walter and her husband opened a remarkable Institut de Beauté on Marszałkowska Street, where Mrs. Walter gave Jewish women lessons on how to appear Aryan and not attract notice.

"There I saw a dozen more or less undressed ladies," Władysław Smólski, a Polish author and member of Zegota, testified after the war. "Some were seated under all kinds of lamps, others, with cream upon their faces, were being subjected to mysterious treatments. As soon as Mrs. Walter came, they all assembled around her, brought up chairs, and sat down, opening books. Then began their catechism instruction!"

Although the women bore Semitic features, each one wore a cross or medallion around her neck, and Mrs. Walter taught them key Christian prayers and how to behave invisibly in church and at ceremonial events. They learned ways to cook and serve pork, prepare traditional Polish dishes, and order the moonshine vodka called *bimber*. Typically, when the police stopped Jews on the street, they checked the men for circumcision and ordered the women to recite the Lord's Prayer and Hail Mary.

The smallest detail could betray them, so Mrs. Walter ran a kind of charm school, the charm of nondetection, which required just the right blend of fashionable makeup, restrained gestures, and Polish folk customs. This meant resisting all Jewish expressions—such as asking "What street are you from?" instead of "What district are you from?" They paid special attention to the habitual and the commonplace—how they walked, gestured, acted in public—with men reminded to remove their hats in church (in temple they would have kept them on), and everyone taught to celebrate their own patron saint's day as well as those of friends and family.

Hair belonged off the forehead, neatly reined in or swept up into more Aryan styles, while bangs, curls, or frizz might raise suspicion. Black hair required bleaching to dull its glitter, but shouldn't become implausibly pale. When it came to choosing clothes, Mrs. Walter advised: "Avoid red, yellow, green or even black. The best color is grey, or else a combination of several inconspicuous colors. You must avoid glasses of the shape that's now fashionable, because they emphasize the semitic features of your nose." And some outstanding semitic noses required "surgical intervention." Fortunately, she worked with Polish surgeons (such as the eminent Dr. Andrzej Trojanowski and his colleagues) who reshaped Jewish noses and operated on Jewish men to restore foreskins, a controversial and clandestine surgery with an ancient tradition.

Throughout history, "reskinning," as the Romans called it, had saved persecuted Jews from discovery, and the Bible reports the practice as early as 168 B.C., during the reign of Antiochus IV, when the Greco-Roman fashion of naked sports events and public bathing emerged in Judea. Jewish men hoping to disguise their lineage had only two choices: they could try to avoid scenes of nakedness, or they could redress their appearance by using a special weight, known as the *Pondus Judaeus,* to stretch the foreskin until it covered the glans. Stretching created small tears between the skin cells, and as new cells formed to bridge the gap, the foreskin lengthened. No doubt this took a while, hurt, and wasn't always easy to hide, though clothes of the era draped loosely. During World War II, the same effect could be achieved surgically, though, needless to say, medical literature of the Nazi era doesn't detail the procedure.

In the circles within circles of Underground life, Jan surely knew the Walters; the bleach and recipe Antonina used may well have come from their salon. Mrs. Walter and her elderly husband hid five Jews at a time in their own home and offered "an endless chain" of people lessons in "good looks" at the Institut de Beauté throughout the war. In later years, Mrs. Walter wrote that "the accidental fact that not one of the casual inhabitants of our war-time nest fell victim to disaster gave rise to a superstitious legend which continually increased the influx of guests." In fact, she explained, her actions were a simple voodoo of compassion: "Suffering took hold of me like a magic spell abolishing all differences between friends and strangers."

Chapter 26

As spring sidled closer and nature hovered between seasons, the snow melted and a low green cityscape of garden plants arose during the day, but at night the land froze over again and moonlight glittered the walks into silver skating trails. Hibernating animals still curled up underground, waiting in suspense. The villa's people and animals sensed the lengthening light, and when a gust of air swept indoors, it carried the mossy sweet smell that rises from living soil. The faint pink coating the treetops promised rippling buds, a sure sign of spring hastening in, right on schedule, and the animal world getting ready for its fiesta of courting and mating, dueling and dancing, suckling and grubbing, costume-making and shedding—in short, the fuzzy, fizzy hoopla of life's ramshackle return.

But spring floated outside the small rupture in time the war had gouged. For people attuned to nature and the changing seasons, especially for farmers or animal-keepers, the war snagged time on barbed wire, forced them to live by mere chronicity, instead of real time, the time of wheat, wolf, and otter.

Confined to her bed's well-padded prison, Antonina rose

occasionally to hobble the few painful steps onto her balcony, from which she had a wide view, and could even hear the powerful noise of ice cracking on the Vistula River, a tympani signaling winter's end. Being bedridden had slowed the world down, given her time to page through memories, and brought a new perspective to some things, while others lay beyond reach or evaded her view. Ryś spent more unsupervised time, but she reckoned him "more capable and levelheaded than any child his age should have to be."

Older children, from youth groups aiding the Underground, had begun arriving unexpectedly, and neither Antonina nor Ryś knew who would be appearing when; though Jan had warning, he was often away at work when they floated in like clouds or just as suddenly vanished. They usually stayed in the Pheasant House for a night or two, then melted back into Warsaw's undergrowth, with Zbyszek, a boy high on the Gestapo's most-wanted list, lingering for weeks. It fell to Ryś, as the least conspicuous villa-ite, to deliver their meals.

Antonina and Jan never spoke of the scouts' doings in front of Ryś, even if some appeared like sightings of rare animals, then mysteriously slipped away, and to her bafflement Ryś didn't seem to care much, despite his usual curiosity. Surely he'd fabricated some story about them? Wondering what, she asked him if he had any thoughts about the young visitors, any opinion about Zbyszek, for instance.

"Oh, *Mom*," Ryś said in the long-suffering tone children reserve for benighted parents, "I know all about it! A *man* can naturally understand these things. I never asked you any questions because I could see that you and Zbyszek had secrets you didn't want to share with me. But I don't care about Zbyszek! I have my own friend now. Anyway, if you really want to know what I think of Zbyszek—I think he's a stupid boy!" And with that Ryś shot out of the room.

Antonina wasn't surprised by his jealousy, which seemed only normal, but Ryś had become more secretive of late, she thought, and much less talkative. Realizing something had collared his attention, she wondered what he was up to. The only answer that rose to mind was his new friend, Jerzyk Topo, the son of a carpenter whose family had recently moved into a staff apartment on the zoo grounds. Antonina found Jerzyk polite and well behaved, a few years older than Ryś, handy with tools, a boy learning his father's trade. Ryś admired his woodworking skills, the two shared an interest in building things, and since they lived close they played together every day. From her second-story watchtower, Antonina sometimes glimpsed them building secret shapes and talking constantly, and she felt relieved that he'd found a new playmate.

Then one day, after the boys had gone to school, Jerzyk's mother appeared at the villa and anxiously asked Antonina if they could talk in private. Antonina ushered her into her bedroom and closed the door. According to Antonina's account, Mrs. Topo then said:

"The boys have no idea I'm here. Don't tell them! I don't really know how to begin. . . ."

Antonina began to worry—what had her son done?

Then Mrs. Topo blurted out: "I was eavesdropping on them—I'm sure they didn't see me. And I know that's a terrible thing to do, but how could I help myself once I got wind of what they were planning? I had to learn what they were up to. So I was quiet and listened, and I was shocked! I didn't know whether to laugh or cry. When they left I didn't know what to do, and I decided I'd better come and talk to you. Maybe together we can figure something out!"

Antonina found the news alarming. Could Mrs. Topo be overreacting to the boys' innocent capers? Hoping so, she said:

"Your son is such a good boy. I'm sure he wouldn't do anything to hurt you. And Ryś is still so little. . . . Okay, I can watch him more closely. . . . But what exactly did our boys *do*?"

"They didn't do anything wrong *yet*, but they're planning something big."

Antonina wrote that "my heart fell into my feet" as Mrs. Topo explained that she'd overheard the boys pledging to oust the Germans, which they believed their patriotic duty, first by hiding a bomb in a tall haystack near the Germans' storehouse of weapons near the zoo fence.

"And under Jerzyk's mattress," Mrs. Topo continued, "I found one of your towels, with big red letters on it that said 'Hitler kaput!' They want to hang this towel above the main gate of the zoo, because there are so many Germans coming here all the time and they're bound to see it! What are we going to do? Maybe your husband could talk to them and explain that they're much too young to fight, and if they go through with their plan they'll put us *all* in danger. . . . But what do you think we should do?"

Antonina listened quietly, trying first to absorb, then analyze the disturbing news that she found both noble and foolish. She assumed Ryś had concocted the idea while eavesdropping on the scouts, who were staging similar acts of sabotage. By now, not drawing attention to the bustle at the zoo had become a fine art, like sleeping with dynamite. All they needed was the boys hanging up a literal red flag.

She also wondered how she could have missed this plot of Ryś's, and misjudged his ability to understand the grown-up world of consequences, when she'd thought she could count on his absolute secrecy, and on her ability to gauge his maturity. Her anger at him and at herself quickly turned to sadness as she realized that

> instead of praising his bravery and initiative, and telling him how proud he made me, I had to punish him, and tell his father that he stole some explosives, and maybe even embarrass him in front of his friend. I knew Jan would be furious.

"Yes," she said to Mrs. Topo, "I will ask Jan to talk with the boys. Meanwhile, it is best to burn the towel."

That evening, she overheard her menfolk, father and son, quietly talking in a formal, military way:

"I hope you appreciate that I'm not treating you like a child, but like a soldier," Jan said, appealing to his son's natural wish to be taken seriously as a grown-up. "I am an *officer* in this house and your *leader*. In the military field of action, you must do *only* what I order, nothing on your own. If you want to continue having this kind of relationship with me, you have to swear that you won't do anything without my knowledge. The action you planned with Jerzyk falls into the category of 'anarchy,' and 'arbitrary'; and you should be punished for it—just like you would be in the regular army."

But what punishment should a father in the role of a military leader impose on a small child in the role of a soldier? Risk isn't shaped the same in a child's eyes, nor can a child see as far downstream from an event, and punishment works only if both parties feel it to be fair, *fairness* being the gold standard of childhood.

So he said: "Maybe you want to suggest *how* I should punish you?"

Ryś considered it seriously. ". . . You can spank me," he finally offered.

And presumably Jan did, because Antonina, in recording the scene in her diary, noted simply: "And, in this small way, our own private family Underground ceased to exist."

Chapter 27

In spring of 1943, Antonina rose from bed at last, in tune with hibernating marmots, bats, hedgehogs, skunks, and dormice. Before the war, she had loved the yammering zoo in springtime, with all its noisy come-ons, *bugger off!*s, and hallelujahs, especially at night, in the quiet city, when feral noises leapt from the zoo as from a giant jukebox. Animal time colliding with city time produced an offbeat rhythm she relished and often wrote of, as in this reverie in her children's book about lynxes, *Rysie*:

> When the spring night wraps Warsaw in a dark coat, and the glaring, luminous signs scatter the dark streets with cheerful reflections, when the quiet of the sleeping city is interrupted by a horn of a late car—on the right bank of the Vistula, among old weeping willows and poplars, the secret sounds of wilderness and the piercing rumble of the jungle is heard. A dance band made up of wolves, hyenas, jackals, and dingoes is heard. The cry of an awakened lion seizes the neighboring population of monkeys with horror. Startled birds scream terrifyingly in the ponds, while in their cage Tofi and Tufa [lynx kittens]

croon a homesick serenade. Their meowing in sharp and penetrating notes rises over the other night sounds of the zoo. Far from the untouched corners of the world, we think about the rule of Mother Nature, with her untold secrets still awaiting discovery, we live among our earthly companions, animals.

While chill still clung to the air and her muscles felt faint from disuse, she lived in a cocoon of woolen undergarments, heavy sweaters, and warm stockings. Wobbling around the house with a cane, she had to learn how to walk again, her knees trembled, and things slipped from her fingers. A toddler again after so many years, she felt cosseted by Magdalena and the others, who allowed her to be a sick little girl, fussed over by family, but she also scolded herself and "felt so embarrassed and useless." For three months, others had done her work for her, waited on her, nursed her, and even now, eager to return to her role in the working household, she couldn't handle chores. "What kind of woman am I?" she chided herself. Whenever she said so in their earshot, Magdalena, Nunia, or Maurycy would counter with:

"Stop that! We're helping you out of pure selfishness. What on earth could we do without you? Your only job is to get stronger. And to give us our orders! We've missed all your energy, wit, and, okay, sometimes your scatterbrained behavior. Amuse us again!"

Then Antonina would laugh, brighten, and slowly wind the machinery of the crazy household as if it were an antique clock. She wrote that they monitored her constantly, fussed, "didn't allow me fatigue, chill, hunger, or worry," and in return, she thanked them for "spoiling me like no one ever had." Writing those words is the closest she comes to talking about being an orphan. Always present in their absence, her dead parents belonged to singed events, a grief before words when she was only nine, a final end at the hands of the Bolsheviks too horrible for a child to keep imagining. They may have haunted her memories, but she never mentions them in her memoirs.

Antonina's friends bundled her up, encouraged her to heal through rest, and embraced by that close circle, she thrived and sometimes "even forgot the occupation" and her "relentless yearning for the war to end soon."

Jan continued to leave the house early and return just before curfew hour, and, although the villa-ites never saw him at work, at home they found him short-tempered and uneasy. To keep their life livable, he checked and rechecked every ritual and routine, a taxing responsibility, since the tiniest chaos, neglect, or impulse could unmask them. Small wonder that he rigidified from the strain and began addressing them as his "soldiers" and Antonina as his "deputy." Jan ruled the villa and the Guests couldn't disobey him, but the atmosphere began to sour because, as a volatile dictator, Jan apparently made daily life tense by often yelling at Antonina, despite her efforts to please him. In her diary, she wrote that "he was always on the alert, took all the responsibilities on his shoulders, and protected us from bad events, trying to check everything very carefully. Sometimes he talked to us as if we were his soldiers. . . . He was cold and expected more from me than from the rest of people in our household. . . . [T]he happy atmosphere in our home was gone."

She went on to say that nothing she did ever seemed good enough, nothing made him proud of her, and perpetually disappointing him felt wretched. In time, her loyal, angry Guests stopped talking to Jan entirely or even making eye contact with him—hating how he treated her but unwilling to confront him, they blotted him out. Jan bristled at their silent protest, complained that civil disobedience in the household wouldn't do, and anyway why were they blaming and excluding *him*?

"Hey, everybody! You're ignoring me just because I criticize Punia a little," he said, using one of his pet names for her (Little Wildcat, or Bush Kitten). This is undeserved! You think I don't have a say here at home? Punia isn't always right!"

"You're away all day," Maurycy said quietly. "I know your life outside of this house is full of all sorts of dangers and traps. But that keeps it interesting, too. Tola's situation is different," he said, using another of Antonina's nicknames. "It reminds me of a soldier who's on constant duty on a battlefield. She has to stay alert all the time. How can you not understand this and scold her for being a little absentminded now and then?"

One afternoon in March the housekeeper yelled from the kitchen, "My God! Fire! Fire!" Looking out the window, Antonina saw a huge mushroom of smoke and flames, a devouring blaze in the Germans' storage area, where a blast of wind was spreading fire like honey across the roof of the barracks. Antonina grabbed her fur coat and ran outside to check on the zoo buildings and the fox farm, which stood only a gust away from the flames.

A German soldier biked up fast to the villa, dismounted, and said angrily:

"You set this fire! Who lives here?"

Antonina looked at his hard face and smiled. "You don't know?" she said pleasantly. "The director of the old Warsaw Zoo lives here. I am his wife. And we're much too serious for pranks like setting a fire."

Anger met by pleasantries is hard to sustain, and the soldier calmed down.

"Okay, but those buildings over there—"

"Yes. Our previous employees occupy two small apartments. They're nice people I know and trust. I'm sure they didn't do it. Why would they risk their lives to burn down a stupid haystack?"

"Well, *something* started it," he said. "It wasn't lightning. Someone had to *set* this fire!"

Antonina faced him innocently. "You don't know? I'm almost positive who set the fire," she said.

The amazed-looking German waited for her to solve the mystery.

As Antonina continued in a friendly, conversational tone, rarely used German words floated up from a deep bog of memory. "Your soldiers take their girlfriends to that place all the time. The days are still pretty cold, and it's cozy to sit in hay. Most likely a couple was there again today, they smoked a cigarette, and left a butt there . . . and you know the rest." Despite her poor German, he understood perfectly well and started to laugh.

Heading into the house, they talked of other things.

"What happened to the zoo's animals?" he asked. "You had the twelfth elephant born in captivity. I read about it in the newspaper. Where is she now?"

Antonina explained that Tuzinka survived the first days of bombings, and that Lutz Heck had shipped her to Königsberg with some other animals. As they approached the porch, two German policemen pulled up on a motorcycle with a sidecar, and her companion told them the whole story, after which the men laughed crudely, then they all went indoors to write a report.

Soon after they left the telephone rang. She heard a stern German voice say, "This is the Gestapo," then speak too fast for her to follow. But she caught the words "Fire?" and "Who am I talking to?"

"The haystack was on fire," she said as best she could. "A building burned down, a fire truck came, and everything is fine now. The German police were already here and they wrote a report."

"You say they did an investigation? Everything is fine? Okay. *Danke schön.*"

Her hand was shaking so hard she had trouble setting the telephone receiver back on its cradle, as all the events of the past hour started flooding in on her and she replayed them inside her head, making sure she'd done and said the right things. With the coast clear, the Guests came out of hiding and hugged her, praising her bravery. In her diary, she noted that she "couldn't wait to tell Jan."

During supper, Jan listened to the whole story, but instead of the hoped-for approval, he grew quiet and thoughtful.

"We all know that our Punia is a wunderkind," he said. "But I'm a little surprised that everyone is so excited about this event. She acted exactly the way I'd expect her to. Let me explain what I mean from a psychological point of view.

"You already know from our stories about the zoo before the war that whenever I had a tough problem with some animal—whether it was sick or hard to feed or just too wild—I would always assign the animal to Punia. And I was right to, because nobody can handle animals as well. Why am I telling you this? Not as an advertisement for her, or to prove how wonderful she is, or how much in love I am, or to make her feel good. As we all know, even as a child, Punia lived around a lot of animals and identified with them.

"It's as if she's porous. She's almost able to read their minds. It's a snap for her to find out what's bothering her animal friends. Maybe because she treats them like people. But you've seen her. At a moment's notice, she can lose her *Homo sapiens* nature and transform herself into a panther, badger, or muskrat!"

"Well, as an artist working with animals," Magdalena said, laughing, "I have a flawless eye for these things, and I've always said that she is a young female lion."

Jan continued: "She has a precise and very special gift, a way of observing and understanding animals that's rare, certainly not typical for an untrained woman naturalist. It's unique, a sixth sense."

Antonina listened with pride to her husband's surprising speech, a banquet of praise so lengthy and rare that immediately afterward she recorded his words verbatim in her diary, adding: "He was talking about my talents, praising me in the presence of other people. It never happened before! . . . He was serious!? He had called me 'silly' so often I'd started hearing it as a second name."

"I'm talking about this," Jan said, "to explain a little how animals react in different situations. We know how cautious wild animals can be, how easily they scare when their instinct tells them to defend themselves. When they sense a stranger crossing their territory, they get aggressive for their own protection. But, in Punia's case, it's like that instinct is absent, leaving her unafraid of either two- or four-legged animals. Nor does she convey fear. That combination might persuade people or animals around her not to attack. Especially animals, which are better at telepathy than humans, and can read each other's brain waves.

"When our Punia radiates a calm and friendly interest in her animals . . . she works as a sort of lighting rod for their fear, absorbs it, neutralizes it. Through her comforting tone of voice, her gentle movements, the safe way her eyes meet them, she imparts a trust in her ability to protect them, heal them, nourish them, and so on.

"You see what I'm trying to say—Punia is able to emit waves of calm and understanding. Humans aren't as sensitive as other animals when it comes to signaling in this way, but everyone can tap into some of these invisible waves, more or less, depending on how sensitive their nervous system is. I think some people are much better at catching these signals, and I don't think that it has any connection with intellectual ability. It may even be that more primitive organisms are more receptive. If we were to use scientific nomenclature, we might ask: What kind of psychic transmitter is Punia, and what kind of message is she sending?"

Jan seems to have been influenced by Friedrich Bernhard Marby (1882–1966), an occultist, astrologer, and anti-Nazi who combined the occult tradition of Nordic runes with the scientific principles of his day:

man as a sensitive receiver and transmitter of cosmic waves and rays, which animated the entire universe and whose specific nature and effect were dependent on planetary influences, earth magnetism, and the physical form of the landscape.

If Jan were alive today, he'd know about the role of *mirror neurons* in the brain, special cells in the premotor cortex that fire right before a person reaches for a rock, steps forward, turns away, begins to smile. Amazingly, the same neurons fire whether we do something or watch someone else do the same thing, and both summon similar feelings. Learning from our own mishaps isn't as safe as learning from someone else's, which helps us decipher the world of intentions, making our social whirl possible. The brain evolved clever ways to spy or eavesdrop on risk, to fathom another's joy or pain quickly, as detailed sensations, without resorting to words. We feel what we see, we experience others as self.

"It's a funny thing," Jan went on, "she's not a child, she's not stupid, but her relationship with other people tends to be very naïve; she believes that everyone is honest and kind. Punia knows that there are bad people around her, too, she recognizes them from a distance. But she really can't believe that they may hurt her.

"Another thing Punia has going for her is the way she observes her surroundings and notices every little detail. She saw German soldiers dating their girls on that haystack, and knowing the Germans' crude sense of humor, she used it well in this particular situation. She didn't worry about her German vocabulary being poor, because her voice and speech are very musical and calming. Her instinct and intuition told her exactly what to do. And, of course, her looks were her trump card. She's tall, thin, blond—the ideal figure of a German woman, the Nordic type. I'm sure that was a big plus, too.

"But if you want to know what I think about the *outcome* of this tragic comedy, I think the Germans found Punia's explanation for the fire that destroyed their buildings very *convenient*. It gave them an excuse not to investigate all the stealing that's been going on over there. The fire was an easy way to cover up crime. If they *really* wanted to punish someone, Punia wouldn't have had such an easy time.

"I don't want to criticize your heroine—Punia did a great job. She was very clever, and I'm glad I can trust her, but I like to look at things from a more cynical point of view."

He'd made her near nightmare sound relatively unimportant, her response cool and calculating, maybe as he imagined it would have been for him. Talented and omnicompetent as Antonina was, she revered and deferred to Jan, often felt inadequate, and was perpetually trying to live up to his expectations and gain his approval. At times Ryś, following his father's example, snarled that as a male even *he* could understand things beyond a ditzy female. Yet Antonina comes across from her diaries as someone who felt deeply loved by Jan, Ryś, and the Guests, and an important complement to her husband, whom she regarded as strict with everyone, most of all himself. She also agreed with him about the subtle ways that all animals communicate. After Jan's mini-lecture on her mind-bending, she found it hard to sleep. Such praise in front of her friends! Rare as light in a Polish winter.

"Jan was right, the German soldiers' reaction to my

telepathic waves was similar to the zoo animals," Antonina reflected in her diary. There were many mystical episodes in her past when she felt certain she could build an invisible bridge with animals, make them listen to her requests, bridle their fear, trust her. According to Antonina, her first experiences of this sort happened when she was a girl and spent all her free time in the stables around high-spirited, purebred horses, but for as long as she could remember animals had calmed down around her. Maybe her unusual degree of empathy and alert senses were part of a more creatural sensibility some people inherit, one tinted and tilted by childhood experience. Also, and importantly in Antonina's case, children with insecure attachments to their parents sometimes forge a strong bond with nature itself.

That night she lay awake thinking about the thin veil between humans and other animals, only the faintest border, which people nonetheless drew as "a symbolic Chinese Wall," one that she, on the other hand, saw as shimmery, nearly invisible. "If not, why do we humanize animals and animalize humans?"

For hours, Antonina lay thinking about people and animals, and how little animal psychology had developed compared with other sciences, say chemistry or physics. "We're still walking, eyes closed, in the labyrinth of psychological enigma," she thought. "But, who knows, maybe one day we'll discover the secrets of animal behavior, and maybe one day we'll master our bleaker instincts."

Meanwhile, Antonina and Jan ran their own informal study throughout the war, living closely with mammals, reptiles, insects, birds, and an arcade of humans. Why was it, she asked herself, that "animals can sometimes subdue their predatory ways in only a few months, while humans, despite centuries of refinement, can quickly grow more savage than any beast"?

Chapter 28

1943

As safety ebbed and flowed during the war, even a quiet offhand remark could trigger a landslide of trouble. Word filtered back to Antonina and Jan that one of their Polish zoo guards had caught sight of Magdalena and gossiped that the famous sculptor hid in the villa. Though Antonina judged the guard "decent, maybe even kindhearted—after all, he hadn't called the Gestapo"—she worried lest a careless word reach the wrong ears and the villa's house of cards collapse. "Did the Gestapo already know?" she asked herself. "Was it only a matter of days?"

Grand and petty blackmail, rife in Warsaw, also posed a disabling threat. Thanks in part to the popularity of the black market before the war, and the familiar ease of smoothing one's way with small tips and bribes, Warsaw had swiftly become a city crowded with predators and prey of all sizes, including the decent and bribable, the indecently unbribable, a hard-core criminal element, opportunistic denizens, people hobbled by fear, Nazi sympathizers, and risk-takers who juggled their own and other lives like lit torches. So for the time being it seemed wisest to hide the Guests elsewhere. Mrs. Dewitzowa, who had taught school with Jan before the

232

war, offered Magdalena and Maurycy space in her suburban home; but after only a few weeks, frightened, she sent them back, claiming that suspicious strangers had begun watching her house. Antonina wasn't so sure. "Could the suburbs be riskier than Warsaw?" she wondered. Maybe so, but she suspected something subtler than that, a symptom of how people handle living with fear and uncertainty.

Emanuel Ringelblum wrote of a "psychosis of fear" that many people felt about escaping to the Aryan side:

> It is the imaginary perils, [the] supposed observation by the neighbour, porter, manager, or passer-by in the street, that constitute the main danger, because the Jew . . . gives himself away by looking around in every direction to see if anyone is watching him, by the nervous expression on his face, by the frightened look of a hunted animal, smelling danger of some kind everywhere.

Even if to others Antonina often appeared calm, her writings reveal a woman often assailed by worry and broadsided by fear. She knew the impression she created, as the villa's ballast, and she insisted that "the warm, friendly, almost therapeutic" atmosphere of the villa implied a safety that was only illusory. True, the villa provided a spacious environment for Guests, who weren't forced to live crippled up behind walls or crowded and damp underground. But as the Nazis imposed a tighter choke hold, the game of diverting eyes and cheating death became the art of making possibilities materialize and watching for signs. According to Polish folklore:

A picture falling off a wall, crunching noises beneath a
window, a broom falling without cause, a clock ticking
where there was none. A table that made cracking noises,
a door that opened by itself—all foretold death drawing
near.

To achieve safety brought many inconveniences, such as
having to shop often and buy things in small quantities so
as not to attract too much notice, or drying some clothes
indoors because one didn't dare hang out laundry that couldn't
belong to anyone in the house. Inevitably, fear raided everyone's mood. But as zookeepers, the Żabińskis understood both
vigilance and predators; in a swamp of vipers, one planned
every footstep. Shaped by the gravity of wartime, it wasn't
always clear who or what could be considered outside or
inside, loyal or turncoat, predator or prey.

At first, no one knew about the zoo's custodial secret and
they had to find extra food and improvise escapes entirely
on their own. Luckily, they discovered that an old friend,
Janina Buchholtz, a psychologist and devotee of the arts, was
a key member of Zegota. During the occupation, Janina officially worked as a registered translator for the public notary,
the office where Antonina had stopped for news after visiting
the bombed-out zoo in 1939. Because she handled many
documents, applications, and petitions, papers spilled from
tables, mounded on shelves, rose in precarious stalagmites
on the floor, and seemed ready to cascade everywhere. A
bureaucrat's nightmare, the clutter camouflaged the office's
real life as an Underground nerve center where Aryan documents were forged, safe apartments sought, messengers

dispatched, cash distributed, sabotage planned, and letters sent to people in other ghettos. Contacts received their instructions and filed reports in her office, which meant lots of foot traffic, but like the Żabińskis, she perfected the art of hiding things in plain sight, in this case among enough clutter to make snooping Nazis recoil, reluctant to sift through the dusty, never quite stationary piles. As one survivor recalled, the Nazis "aimed step by step, by means of interlocking decrees, to create a reporting and documentation system that would render any kind of machinations impossible and that would locate every single inhabitant of the city with appropriate precision." This necessitated elaborate false identities, documents, and provenances for people in hiding, because Polish Catholics, who mainly lived in housing blocks, could produce church and municipal records from before the war, including birth, baptism, marriage, tax, death, and inheritance documents. Fresh documents sometimes meant "solid" papers that could withstand Gestapo delving and sometimes flimsy ones (called *lipne*, from the word for linden, which later evolved into "white lies"), which wouldn't pass much scrutiny. As Gunnar Paulsson relates the process,

Setting yourself up as a *homo novus* required not only the creation of a new identity but the severing of all ties to the old, tainted ones. Therefore you had to move. Your former self could then vanish, while the new self registered in the normal way in the new quarters. . . . [Y]ou had to de-register at the registry office in the old district, receiving a coupon in return. You then registered with the building manager at the new place and again

received a coupon. Both coupons then had to be taken
to the local registry office, within a certain grace period,
as proof of registration. . . . [T]o break the chain of
evidence, it was necessary to have a forged de-registration
coupon, and this needed to have backing in the files of
the registry office.

Fortunately, Janina worked in the registry office, where she
could confect identities and plant records to back them up.
Some people claimed to have been born in the Soviet Union,
or to Muslim Poles, or to have lost their papers in a church
fire before 1939; others assumed the identity of a citizen
living abroad or dead. All of these required forgery and finesse,
generating, planting, and altering the records in long chains
of evidence—hence the paper Alps in her office. In 1941,
when Hans Frank decreed that identity cards (*Kennkarte*) be
issued, complete with serial number and fingerprints, clerks
managed to stall them until 1943, and then use the occasion
to make fictitious identity cards. Hordes of people seemed
suddenly to have lost their records. Greedy opportunists and
Underground specialists alike crafted so many passports and
other documents that by the summer of 1943 even Ziegler's
office estimated 15 percent of all identity cards and 25 percent
of all working papers had been forged. One cell of Zegota
alone is credited with generating fifty to one hundred documents a day, papers that ran the gamut from birth and death
certificates to the IDs of low-level SS and Gestapo officials.
Janina pictured her clients as people "walking on quicksand."

"I am lucky . . . I can do wonders," she proudly told her
friend and colleague Barbara "Basia" Temkin-Berman, while

smiling and tapping a crooked finger on the café table to ward off bad luck.

Tall, heavyset, and elderly, Janina always wore the robelike black skirts of a prioress and a peculiar little veiled hat tied under her chin, and she carried a muff. Spectacles balanced on her long thin nose, over eyes brimming with such warmth that people typically referred to her as "the kindest person I've ever known," or "the perennial protector of the underdog."

In her double conspiracy of fighting the Germans and helping the Jews, Janina worked closely with Basia, a psychologist before the war and her physical opposite: a small, slender, nervous, volatile woman who always wore an old wine-colored coat, black beret, and veil to hide her Semitic features.

Janina and Basia conferred daily at the Miodowa Street office or at the *cat*-safe café at 24 Miodowa Street, and together they forged contacts among nuns and priests, railway workers, professors, market-stall owners, shopkeepers, maids, trolley drivers, farmers, beauticians, engineers, clerks and secretaries (willing to erase people from public files or issue bogus certificates). And, of course, a zoo director and his wife. One day Janina spoke with Underground leaders about Magdalena's risk at the zoo, and their decree, though disturbing, made sense to Antonina and Jan. Maurycy would remain in the villa and Magdalena would lodge with an engineer friend of Janina's who lived in Saska Kępa, on the east side of the river, in a lovely old parish complete with a park inhabited by statues: *The Dancer, Rhythm,* and nakedly voluptuous *Bather.* It was a district of neoclassical public buildings, newly built modernist houses with lots of shrubbery, and avant-garde villas of concrete and glass designed between the wars.

At first, the zoo had served only as a temporary shelter, one whistle stop in an elaborate Underground railroad, and Jan and Antonina hid only friends and acquaintances, but later, working with Janina, "things became more organized," as Jan put it in his understated way, by which he meant that, with the Underground's help, he amplified his efforts and began taking unearthly chances.

Of all the Guests to leave the villa, "high-spirited Magdalena, full of energy and laughter," was the one Antonina said she missed most. The two shared an extraordinary friendship, girlish and mature, intimate and professional. Both Jan and Antonina admired Magdalena as an artist, but they also treasured her as a buoyant, funny, generous friend. According to Antonina, losing Magdalena felt physically painful, even though her departure made room for another Guest, another life saved. Jan and Antonina promised to visit Magdalena in Saska Kępa as often as possible; and Maurycy, who couldn't travel safely through the city, wondered if their goodbye meant months, years, or forever, and "took it especially hard."

By late June of 1943, Jan and Antonina figured no one had reported them to the Gestapo, and gingerly began accepting Guests once again. Janina sent them a young friend of hers, Aniela Dobrucka, who had "good looks," as locals said, meaning strong Aryan features, which had allowed her to spend days as a street seller of bread and croissants and nights lodging with an eccentric old woman. Antonina liked the spunky young woman with dark hair, sea-blue eyes, and a temperament both "sweet-natured and a bit mischievous." Coming to Warsaw from a poor farming village, Aniela had struggled to pay her own way through Lwów University.

Rachela Auerbach was her real name, but that became buried in Underground life, where identities melted and one assumed fresh names, guises, and tasks, as required.

Polish émigré Eva Hoffman writes movingly about the psychic earthquake of having to shed her name: "Nothing much has happened, except a small, seismic mental shift. The twist in our names takes them a tiny distance from us—but it is a gap into which the infinite hobgoblin of abstraction enters." Suddenly her given name and her sister's no longer existed, even though "they were as surely us as our eyes or hands." And the new names were "identification tags, disembodied signs pointing to objects that happen to be my sister and myself. We walk to our seats, into a roomful of unknown faces, with names that make us strangers to ourselves."

Lucky enough to leave the Ghetto before the worst of times, Aniela dedicated her life to scrounging food for famished people and working at a hospital and a library, and she was one of the elite few who knew the secret of the milk churns. In the section of the Ghetto devoted to workshops, the OBW (Ostdeutsche Bauwerkstütte) carpentry shop was managed by Germans who obliged the original Jewish owners to continue running daily operations. One of those brothers, Alexander Landau, belonged to the Underground and hired many of its members, supposedly as trained craftsmen, though their lack of basic carpentry skills wasn't always easy to hide. The Halmann carpentry shop, at 68 Nowolipki Street, employed other so-called carpenters, and the houses assigned to them became the center of the Jewish Fighting Organization. Together, these two workshops, by employing many people, kept them from deportation, housed others on the run, and became sites of schools and the hub of much Underground activity.

Only a month after the Germans occupied Poland, historian Ringelblum conceived the idea of an archive, because he felt what was happening was unprecedented in human history, and someone should accurately report the facts and bear witness to the unspeakable suffering and cruelty. Aniela aided Ringelblum with the archives, though Janina read the documents first and hid them for a while inside the upholstery of the big couch in her office. Then this secret group of archivists, code name *Oneg Shabat* (because they met on Saturdays), hid the documents in boxes and milk churns under the Halmann workshop. In 1946, survivors combing through the ruins of the Ghetto found all but one of the milk churns, filled with vividly detailed accounts written in Yiddish, Polish, or Hebrew, which now reside in the Jewish Institute in Warsaw.

In time, Aniela brought to the villa her friend Genia (Eugienia Sylkes), who had organized secret schools in the Ghetto, fought in the Underground Army, and helped plan the Ghetto Uprising. Ultimately captured and forced to board a train for Treblinka, she and her husband jumped from the train near Otwock as it slowed down at a siding to let another train pass (some cars had small windows strung with barbed wire that could be cut or doors that could be forced). In a postwar interview with London's *White Eagle-Mermaid*, Genia recalled that after the jump

I was deadly tired and hungry, but I was too afraid to get closer to the buildings. . . . I couldn't find my husband

and very slowly, using side roads, I went to Lublin. After two days I decided to go back to Warsaw. I traveled with blue collar workers and reached the Old City by early morning. My cousin, the wife of a Pole, was hiding with a Mrs. Kowalska. I went there and was welcomed as a ghost from the other world, I got food, took a bath and went to bed. After a few days, when I was on my feet again, they gave me clothes and I went to Miodowa Street #1, to Janina Buchholtz of Zegota. There I got documents and money. Later my cousin's husband found a room for me at Chłodna Street in a Polish policeman's apartment. I can only speak about all these people who helped me with the highest admiration and affection.

When the policeman's apartment grew unsafe, Janina brought her to the zoo, where, officially, she served as Antonina's tailor who repaired clothes and later, when Antonina was pregnant, sewed diapers and baby clothes. Tall and Aryan-looking with a short snub nose, she might have passed easily, but knew little Polish, so in public she pretended to be mute, or sometimes Estonian, as her false papers declared. Feigning mutism, she joined a cadre of other heavily accented people floating around the city, silenced by the unspeakable.

Chapter 29

Soon after the bluebells faded in spring, wild garlic clusters grew in the damp shade of old trees, with tiny white flowers oozing a sweet vapor that poured through open windows at dusk, their leaves towering over two feet high in a scramble for light. Some farmers grazed sheep in garlic groves to scent the meat, and others cursed if their cows wandered in by mistake and browsed garlic, tainting the milk. Locals used wild garlic in rejuvenative potions and poultices to lower a high fever, warm fading ardor, dry acne, tune the heart, or ease whooping cough. They bruised the bulbs for cooking, and simmered a wild garlic soup.

"The zoo became immersed in a warm May night," Antonina wrote, sketching the scene in her diary: "Trees and shrubs, house and terrace were flooded by pale aquamarine, a cool and impassive moonlight. Branches of the lilac bushes bent low with heavy, faded clusters of flowers. The sharp, geometric outlines of sidewalks were highlighted by long black shadows. Nightingales sang their spring songs over and over, intoxicated by their own voices."

The villa-ites sat listening to Fox Man's piano concert, losing time and reality in a world lit by candle shadows and the constellations of notes hovering in darkness. "The silent romantic night swelled with the impetuous chords of Chopin's Etude in C Minor. The music spoke to us of sorrow, fear, and terror, as it floated around the room and through an open window," Antonina recalled.

Suddenly she heard a soft uncanny rustling coming from the bed of tall hollyhocks beside the window, a noise she alone seemed to decode between notes. When an owl screeched, warning something or someone away from its nest of fledglings, Antonina read the sign and discreetly told Jan, who went outside to investigate. Reappearing in the doorway, he gestured for her to join him.

"I need the key to the Pheasant House," he whispered. As housewife, she kept the keys, and there were many: some to doors at the villa, others to zoo buildings, still others to doors that once existed, and some that served no memorable purpose but nonetheless couldn't be tossed out. This key would have come easily to hand because they used it often—unlocking the Pheasant House usually meant a new Guest.

Silently questioning with her eyes, Antonina gave Jan the key, and together they went outside, where they caught sight of two boys diving for cover behind some bushes. Jan whispered that these members of the Underground's sabotage wing had set fire to German gas tanks and urgently needed to lie low. They'd been told to run for the zoo, and, unbeknownst to Antonina, Jan had been expecting them all evening with mounting worry. Recognizing their hosts, the boys suddenly stepped into view.

"For several hours we hid in the bushes next to the house,

because we could hear German language being spoken," one said.

Jan explained that the lovely weather appealed to military policemen who visited the zoo for long walks, and several had left only about twenty minutes earlier. With the coast clear now, they needed to hurry inside the Pheasant House. Because pheasants were delicacies, a Pheasant House sounded quite grand to the boys, and one teased: "We'll pretend we're a rare species, right, Mr. Lieutenant?"

"It's nothing special!" Jan warned him. "Not luxurious quarters by a long shot. Only rabbits live in the building now. It's located close to our house, where we can keep an eye on you and bring you food. But I must remind you: from daybreak on you have to practice the silence of the tomb!" He said sternly: "Don't talk or smoke. I don't want to hear any noise coming from there! Is that understood?"

"Understood, sir!" they said.

Silence reigned, the jacket of silence one sometimes finds on a still, moonless night. The only sound Antonina heard was a key clicking in a lock hidden beneath the wild vines on the Pheasant House.

The next morning, when Ryś took Wicek into the garden and strolled toward the aviary, Antonina watched him pause to pet Wicek's long ears and say:

"Be ready now, you old horse! We're going to the Pheasant House! So, remember: Be very quiet!" He raised a shushing finger to his lips. Together they made their way to the small wooden building, with Wicek at Ryś's heel.

Inside, Ryś found two boys sleeping on beds of hay, surrounded by rabbits of all sizes, which, like trolls, were busily watching and sniffing at the sleeping humans. Ryś locked the door behind him, quietly set a basket of milkweed on the floor, and tossed around handfuls of the pods and stems for the rabbits to eat. Then he took out a pot of milk with noodles in it, a big chunk of bread, and two spoons.

Studying them as they slept was irresistible for a boy curious about animals and humans, so he edged his face close to theirs and considered how best to wake them, since he wasn't supposed to stomp, clap, or yell. Squatting, he tugged one boy's sleeve, which didn't stir the exhausted sleeper, then he tugged harder and harder, and still the boy slept. Since the hands-on technique didn't work, he tried something subtler: filling his lungs with air, he puffed at the boy's face until at last he lifted his hand to swat an imaginary insect and his eyes finally opened.

Half conscious and startled, the boy looked panicky, and Ryś decided it might be time to introduce himself, so, leaning even closer, he whispered:

"I am Ryś!"

"Pleased to meet you," the boy whispered back, then added emphatically: "I am *Pheasant*!"

This was an understandable confusion, inasmuch as Ryś is the word for lynx and people hiding at the zoo were given the name of the animal that once lived in their hiding place.

"Yes, but I'm telling the truth," Ryś insisted, "I really *am* Ryś, it's not a joke. I mean Ryś the boy, not Ryś with little crests on its ears and a fox terrier's tail!"

"Yes, I see that," the boy said. "I'm only a pheasant today. Anyway, if you were a real lynx and I had feathers, you'd be eating me by now, right?"

"Maybe not," Ryś said seriously. "Please don't joke. . . . I brought you breakfast and a pencil, and—" Suddenly they heard footsteps on a sidewalk nearby and at least two German voices. Ryś and the boy sat twig-still.

After the voices had passed, Ryś said: "Maybe they're only people heading for their garden plots." The second conspirator woke up, stretched, and massaged his stiff, cramped legs as his comrade showed him the bowl with soup and handed him a spoon. Still squatting, Ryś watched them eat, waiting until they were finished, then said quietly: "Goodbye. Don't get bored. I'll bring you dinner and something to read. . . . You'll get some daylight through the little skylight window."

As he left, Ryś heard one boy say to the other, "Nice kid, isn't he? And it's so funny having a lynx guard the pheasants. It would make a good fairy tale, wouldn't it?"

Ryś returned to the villa and told Antonina all about his adventures with the boys, who hid in the Pheasant House for three weeks while Ryś tended them as his charges, until the Germans gave up searching, new documents were forged, and another hiding place secured. One morning Ryś found nothing but rabbits in the Pheasant House and realized the boys had moved on, which he took personally, as friends abandoning him.

"Mom, where *are* they?" he asked. "Why did they leave? Didn't they like being here?"

She explained that they had to leave, that war wasn't a game, and that other Guests would arrive to fill the void they left.

"You can still take care of your animals," she said, trying to comfort him.

"I prefer *pheasants*," he whined. "Don't you understand—it was different! They even called me 'friend,' and they didn't think of me as a little boy, but as their guardian."

Antonina caressed his blond hair. "You're right," she said, "this time you were a big boy, and you helped in a very important way. You do understand that it's a secret and you mustn't tell anybody about it, right?"

She saw anger jump in his eyes. "I know that better than you do!" he snapped. "These things aren't for *women*," he said contemptuously, then whistled for Wicek. All she could do was watch sadly as the two disappeared into the bushes, knowing Ryś had to cope with yet another abandonment and another secret he could never tell. "If I maintain my silence about my secret it is my prisoner," Gdańsk-born philosopher Arthur Schopenhauer wrote in an earlier era, but "if I let it slip from my tongue, I am *its* prisoner." Recording the day's events in her journal allowed Antonina to juggle secret-keeping and secret-telling—one substance, like water, that merely assumed different shapes.

Chapter 30

1943

During summer, the black fly's Mardi Gras, clouds of insects tormented the zoo as usual and anyone abroad at sunset wore long sleeves and pants, despite the heat. Inside the villa, Wicek the rabbit, on the prowl for something edible, heard a squeaky noise and hopped toward the kitchen, where he found Kuba the chick eating. During dinner, Kuba sometimes roamed the table, pecking up crumbs, with Wicek watching from a distance until, in one broad leap, he would suddenly appear next to a chunk of bread or a bowl of potatoes and start gobbling, to the fright of the chick and the great amusement of the humans.

Whenever Ryś lay awake after curfew, waiting for his father to return home, rabbit and chick perched on the edge of his comforter and sat vigil with him. According to Antonina, at the sound of the doorbell all three would grow excited and listen for Jan's footsteps on the hall stairs, which echoed hollowly, because the wooden staircase floated right above steps leading from the kitchen to the basement, and the space resounded like a muffled drum.

Ryś would search his father's face for exhaustion or worry, and sometimes Jan's chilly hands unwrapped food he'd bought with food stamps, or he returned with an exciting story, or pulled another animal from his magic backpack. After Ryś fell asleep, Jan would quietly head downstairs as the rabbit hopped off the bed, the chick slid down the comforter, and both animals followed him to the dining room table for his evening meal. According to Antonina, the rabbit inevitably jumped onto Jan's lap and the chick crawled onto Jan's arm, then climbed to his neck, where he curled up in the jacket collar and slept; and even when Antonina removed Jan's dishes and replaced them with papers and books, the animals refused to leave the warmth of lap and collar.

Warsaw endured a brutally cold winter in 1943. Ryś caught a bad chest cold that sharpened into pneumonia, and he remained in hospital for several weeks, recovering without the punch of heavyweight antibiotics. Penicillin wasn't discovered until 1939, in war-bound Britain, which couldn't spare scientists to hunt the most fecund mold for human trials. But on July 9, 1941, Howard Florey and Norman Heatley flew to the United States on a plane with blacked-out windows, bearing a small priceless box of penicillin, and joined a lab in Peoria, Illinois, where they studied luxuriant molds from all over the world, only to discover that a strain of penicillin from a moldy cantaloupe in a Peoria market, when submerged in a deep vat and allowed to ferment, yielded ten times as much penicillin as competing molds. The requisite trials finally proved the drug's value as the best antibacterial agent of its day, but wounded soldiers didn't start receiving it until D-Day (June 6, 1944), and civilians and animals not until the end of the war.

When Ryś finally returned home, ice and snow had already started melting from the spring garden, and he could help weed, delve, and plant, with Wicek (whose fur had changed from black to silver gray in winter) hopping beside him, stride for stride like a well-trained dog. The nearly fledged chick pecked at the freshly turned soil, pulling up fat pink worms, and Antonina noted that the *real* chickens, the ones roosting in the chicken house, treated Kuba like an outsider, pecking him fiercely. However, Wicek allowed the chick to climb onto his back and nest deep, and she often saw them hopping around the garden together, rider and steed.

Before the war, the zoo had undulated from one landscape to another—mountains, valleys, ponds, lakes, pools, and woods—depending on the needs of its animals and Jan's fancy as zoo director. But now that the zoo fell under the Warsaw Parks and Gardens Department, Jan answered to a bureaucrat who envisioned one continuous pullulation of green, with every woodlet, hedge, or obelisk echoing the others, according to his design. For that he needed Praski Park and, especially, the zoo's large lawns and arboretum.

In the spring, Director Müller of the Königsberg Zoo, hearing that the Warsaw Zoo had been destroyed, offered to buy all the usable cages. His zoo, though considerably smaller than Jan's, nestled in a fortress city founded by Teutonic knights and thought to be impregnable. Late in the war, Churchill would target Königsberg for one of the RAF's controversial "terror raids," ultimately destroying most of the city (zoo included), which finally surrendered to the Red Army on April 9, 1945, when it became known as Kaliningrad.

But in 1943, as self-crowned "Father of Warsaw," Danglu

Leist, the German president, didn't want his city to be outshone by a smaller one, and he decided Warsaw should have its zoo once more. Antonina described Jan as "ecstatic" when Leist invited him to submit a budget for a reborn zoo, remarking that even with the zoo animals gone, the buildings destroyed, the equipment dilapidated, the zoo still prospered in Jan's heart and imaginings. At last, "phoenix-like," she thought, the zoo, his career, and his passion for zookeeping might flourish again; and his Underground work could only profit from the bustle of an open zoo's daily life, with its moving mosaic of visitors, animals, and workmen, against which the villa's escapades might fade. A restored zoo would vitalize every contour of their life; it was perfect. Too perfect, Jan felt. He began immediately analyzing the plan for flaws, foremost that Poles were "boycotting all amusement activities created by the enemy." Normally the zoo offered a wellspring of research and programs, but, fearing a Polish intelligentsia, the Nazis had allowed only elementary schools to stay open; all high schools and universities were banned. With the zoo's teaching role abolished, it could offer only a small gallery of animals, and with food scarce and the city markets empty, how could the zoo justify feeding its animals? What's more, a zoo might hurt the city's economy, Jan reasoned, or expose him to danger if he didn't run it according to German dictates. While such problems seemed insurmountable, Antonina admired Jan's self-sacrifice, which she felt showed "character, bravery, and a realistic mind."

"It's hard to say what would be best for the city or the zoo," Jan told Julian Kulski, Warsaw's Polish vice president. "What if in fifty or a hundred years someone were to read a history of the Warsaw Zoo, re-created by Germans for their pleasure, even though it drained the city? How would you like that footnote to your biography?"

"I live with this sort of dilemma every day," Kulski moaned. "I swear I never would have taken this job if all the people of Warsaw had been killed in 1939 and the Germans had repopulated the city with outsiders. I'm only doing it to serve my people."

During the next two days, Jan carefully crafted a letter to Leist, in which he praised his decision to reopen the zoo and appended a colossal budget for basics the zoo would require. Leist didn't bother with an answer, nor did Jan expect one, but neither did he expect what happened later.

Somehow the director of Parks and Gardens got wind of the proposed rebirth of the zoo, which would have destroyed his unified parks project, and so to foil Jan he sent word to the Germans that Dr. Żabiński's services were no longer needed and his job should be terminated.

Antonina didn't credit this to "antipathy or revenge," but rather an "idée fixe" of leaving his mark on Warsaw's parks. Still, it threatened Jan and his family, because anyone *not needed* by a German employer lost his working papers, which made him eligible for deportation to Germany to drudge in munitions factories. Since the villa belonged to the zoo, the Żabińskis could easily lose their home, many *melinas*, and Jan's small salary. Then what would become of the Guests?

Kulski doctored the complaint against Jan before the Germans could read it, and, instead of losing work, Jan was transferred to the Pedagogic Museum on Jezuicka Street, a sleepy little enclave with only an elderly director, a secretary, and a few guards, whom the Germans seldom bothered. Jan said his job mainly entailed dusting old physical education equipment and preserving zoological and botanical posters lent to schools before the war. That left him more time to scheme with the Underground and teach biology in the "flying university." Jan also kept a part-time job in the Health Department, so with one thing and another, Antonina and Ryś knew that Jan melted away each morning, to face who knew what hazards, and reappeared in the obscure no-man's-land before curfew. Though Antonina didn't realize exactly what he was up to, her mental cameo of Jan was haloed in danger and potential loss, and she tried to banish the naturally-arising mind-theaters in which he was captured or killed. "But I worried about his safety all day long," she confessed.

In addition to building bombs, derailing trains, and poisoning pork sandwiches headed for the German canteen, Jan continued to work with a team of construction people building bunkers and hideouts. Zegota also rented five flats, just for refugees, who had to be regularly supplied with provisions and moved from one safe house to the next.

Officially, as spoken truths, Antonina knew few of Jan's activities; he rarely told Antonina about them, and she rarely asked him to confirm what she suspected. She found it essential *not* to know too much about his warcraft, comrades, or plans. Otherwise, worry would pollute her mood all day and interfere with her equally vital responsibilities. Because many people relied on her for their sustenance and sanity, she "played a sort of hide-and-seek game" with herself, she noted, *pretending* not to know, as Jan's shadow life floated around the edges of her awareness. "When people are constantly on the brink of life and death, it's better to know as little as possible," she told herself. But, without meaning to exactly, one still tends to conjure up scary scenarios, their pathos or salvation, as if one could endure a trauma before it occurred, in small manageable doses, as a sort of inoculation. Are there homeopathic degrees of anguish? With sleights of mind, Antonina half fooled herself enough to endure years of horror and upheaval, but there's a difference between not knowing and choosing not to know what one knows but would rather not face. Both she and Jan continued to keep a small dose of cyanide with them at all times.

When the governor's office phoned one day, summoning Jan, the villa-ites all assumed he'd be arrested, and as panic filled the house, they advised him to run away while he could. "But then who will guard and support everyone?" he asked Antonina, knowing that he would be condemning them to death.

The next morning, as Jan was leaving for the governor's office, after they had said their goodbyes, she whispered the unspeakable: "Do you have your cyanide with you?"

His meeting was called for 9 A.M., and Antonina swore she felt the seconds ticking away inside while going through the motions of household chores. Around 2 P.M., as she was dropping peeled potatoes into a pot, she heard a voice whisper "Punia," and she looked up, pulse skipping, to see Jan standing right in front of her at the open kitchen window. He was smiling.

"Do you know what they wanted?" he asked. "You're not going to believe this. When I got to the governor's office I was taken by car to Konstancin, Governor Fischer's private residence. Apparently, his guards had discovered snakes around the house and in the woods nearby, and they were afraid members of the Underground might have dumped lots of poisonous vipers there to wipe out the German government! Leist told them to contact me as the only person who knew about snakes. Well, I *proved* there weren't any poisonous vipers by catching the snakes by hand!" Then Jan added somberly: "Luckily, I didn't need the cyanide this time."

Before leaving work one afternoon, Jan placed two pistols in his backpack and covered them with a freshly killed rabbit. As he stepped off the trolley at the Veterans Circle stop, he suddenly encountered two German soldiers, one of whom yelled "Hands up!" and ordered him to open his backpack for inspection.

"I'm lost," Jan thought. With disarming casualness, he smiled and said: "How can I open my backpack with my hands up? You'd better check it yourself." A soldier poked around a little inside the backpack and saw the carcass.

"Oh, a rabbit! Maybe for dinner tomorrow?"

"Yes. We have to eat *something*," Jan said, still smiling.

The German said he could put his arms down, and with an *"Also, gehen Sie nach Hause!"* sent him on his way.

Antonina wrote that as she listened to Jan's account of his close call, the veins in her head throbbed so hard she could feel her scalp moving. That Jan seemed amused as he told her the story, joking "about what might have been a tragedy, upset me even more."

Jan confessed to a journalist years later that he had found

such risks alluring, exciting, and felt a soldierly pride in ridding himself of fear and thinking fast in tight spots. "Cool" is how Antonina described him, a compliment. This thread of his personality, so unlike her own, she found admirable, alien, and also humbling, since she couldn't match his feats of bravery. She had had close calls, too, but whereas she ranked Jan's as heroic, she deemed hers merely lucky.

By the winter of 1944, for example, when the city gas lines didn't work well and their second-floor bathroom had no hot water, pregnant Antonina craved the carnal luxury of a steaming bath. On a whim, she telephoned Jan's cousins Marysia and Mikołaj Gutowski, who lived in the borough of Żoliborz, just north of City Center, a pretty neighborhood on the left bank of the Vistula that once belonged to monks who named it Jolie Bord (Beautiful Embankment). At the mention of hot water, just as she'd hoped, her cousin said they had plenty and invited her to spend the night. Antonina rarely left the villa alone, even to visit the butcher, market, or other shops, but this sybaritic rarity tantalized her, so "after getting permission from Jan," she braved the deep snow, February winds, and German soldiers, and went to their house early one evening.

After a long bath, she joined her cousins in a dining room "beautifully decorated with elegant furniture and objects." Glinting light caught her eye: a framed collection of tiny teaspoons silvering on one wall, each decorated with the emblem of a different German city—inexpensive souvenirs Gutowski had collected on prewar trips. After dinner she went to the guest room and fell asleep, but at 4 A.M. she woke to the growl of truck engines just outside the house, and heard Marysia and Mikołaj running to the front window. She followed them and stood in darkness watching trucks with tarp roofs parked at Tucholski Square, surrounded by a huge crowd and German police, with other trucks pulling up. Antonina wrote that as the soldiers kept loading hostages bound for the camps, she anxiously hoped they wouldn't cart her off, too. Deciding not to get involved, she and her cousins returned to bed, but soon a loud pounding on the door summoned Mikołaj downstairs, still in his pajamas, and Antonina worried what her family would do without her help. Suddenly German soldiers stood in the hallway and asked for her documents.

Pointing at Antonina, a soldier asked Mikołaj: "Why isn't this woman registered here?"

"She's my niece, the zookeeper's wife," he explained in fluent German. "She's just spending the night here because their bathroom is broken; she came to take a bath and spend the night—that's all. It's dark and slippery out, not a good time for a pregnant woman to be on the street alone."

As the soldiers continued inspecting the house, they moved slowly from one elegantly furnished room to another, exchanging smiles of pleasure.

"So gemütlich," one said, a word conveying a pleasant cheerfulness. "Back home bombing raids have destroyed our houses."

Antonina noted later that she could well imagine his sorrow. In March, American bombers had dropped 2,000 tons of bombs on Berlin, and in April thousands of planes had jousted over Germany's once-beautiful cities. The soldiers had much *gemütlichkeit* to long for, though the worst still awaited them.

By the end of the war, the Allies would blanket-bomb German cities, including Dresden, historic seat of humanism and architectural splendor.

Antonina stood to one side and quietly watched their faces for signs of trouble as they entered the dining room, where a soldier spied the pageant of German commemorative spoons on the wall. He paused, edged closer, and then his face flashed surprised delight as he drew his friend's attention to the rows of perfectly arranged spoons, each celebrating a different city. The soldier said politely: "Thank you, everything is fine here, we're finished with our inspection. Goodbye!" And they left.

Thinking over events later, Antonina figured all that saved her were sentimental memories and the idea that someone in the house had German roots. Marysia's whim of buying German souvenirs, and displaying them in the folk-art way, had spared them arrest, interrogation, perhaps death. Despite all she chose not to see, Antonina still hid valuable secrets (people, locations, contacts), and so did Mikołaj, a Catholic engineer who, with Zegota's help, sometimes hid Jews.

At last all went to bed, and the next morning Antonina returned home, where the Guests assured her that if she and Jan could escape narrowly so often, they must live "under the influence of a lucky star," not just a crazy one.

By the time spring came, the hibernating zoo began churning with life, trees unfurled new leaves, the ground softened, and many city dwellers arrived, gardening tools in hand, to work their small vegetable plots. The Żabińskis gave refuge to even more desperate Guests, who joined the villa, underfoot and in closets, or crept into small sheds and cages. Their lack of comforts, photographs, and family relics greatly saddened Antonina, who described them in her diary as "people stripped of everything but their lives."

In June, Antonina affirmed life's relentless optimism by giving birth to a little girl named Teresa, who stole center stage despite the global tug-of-war. Ryś was fascinated by the newborn, and Antonina wrote that she fancied herself back in a fairy tale about a baby princess (Jabłonowski Princess Teresa had been born in 1910), for whom gifts arrived each day. A shiny golden wicker crib, a handmade baby quilt, knitted hats and sweaters and socks at a time when wool was hard to find—these seemed "priceless treasures laden with magic spells of protection." One very poor friend had even embroidered cloth diapers with tiny pearl designs. Antonina doted on the tokens, removing them from tissue paper, touching them, admiring them, arranging them on her comforter like icons. Couples were trying not to give birth during the war, given life's uncertainties, and this healthy baby posed a good omen in one of the most superstitious of cultures, especially about childbearing.

According to Polish folklore, for example, a pregnant woman dared not gaze at a cripple or the baby could become crippled, too. Looking into a fire while pregnant supposedly caused red birthmarks, and looking through a keyhole doomed the baby to crossed eyes. If an expectant mother stepped over a rope on the ground or under a clothesline, the umbilical cord would tangle during childbirth. Mothers-to-be should only stare at beautiful vistas, objects, and people, and could produce a happy, sociable child by singing and talking a lot. Craving sour foods foretold a boy, craving sweets a girl. If possible, one should give birth on a lucky day of the week at a lucky hour to guarantee the baby's lifelong good fortune, whereas a sinister day doled out hexes. Although the Virgin Mary blessed Saturdays, when any newborn automatically evaded evil, Sunday's children could blossom into mystics and seers. Superstitious rituals accompanied the saving and drying of the umbilical cord, the first bath, first haircut, first breast-feeding, and so on. Since it marked the end of infancy, weaning held special significance:

The country women had particular times when they thought weaning was to occur. First, it was not done during the time when birds were flying away for the winter, for fear the child would grow up to be wild and take to the forest and woods. If weaning took place during the time when leaves were falling, the child would go bald early on in life. A child was not weaned during harvest time when the grains were being carefully hidden away, or it would become a very secretive individual.

—*Polish Customs, Traditions, and Folklore*

Also, pregnancy should stay hidden as long as possible, and not be divulged, even by the husband, lest a jealous neighbor cast the evil eye on the baby. In Antonina's day, the evil eye, born of envy to sour and begrudge good fortune, worried many Poles, who believed a happy event invited evil and that praising a newborn cast a vicious spell. "What a beautiful baby" became so poisonous that, as antidote, the mother had to counter with: "Oh, it's an *ugly* child," and then spit in disgust. Following similar logic, when a girl got her first period it was customary for her mother to slap her. The dehexing fell mainly to mothers, who saved offspring by forgoing shows of happiness and pride, thus sacrificing what they prized dearly for what they valued most, because the moment one loved something it became eligible for loss. While to Catholics, Satan and his minions always hovered, Jews also ran a daily gauntlet of demons, the best known of which is perhaps the zombie-like *dybbuk*, the spirit of someone who died and has returned to haunt the body of a living person.

On July 10, Antonina finally emerged from bed, to celebrate Teresa's birth at a small christening party. Traditionally one served braided bread and cheese on such an occasion, to dispel evil forces. Out came bacon-stuffed meat preserves, made from the carcasses of crows shot by Germans the previous winter. Fox Man cooked waffles, and Maurycy made a traditional liqueur of honeyed vodka, called *pępkowa* (navel). Of course, in Maurycy's eyes, the occasion required the presence of his hamster, so Piotr joined the table and began collecting crumbs as usual, carefully checking each plate and cup, perking up his head, sniffing around, whiskers twitching, at last discovering the source of a new aroma which spirited sweetly from the empty liqueur glasses. Lifting a honey-scented glass in his tiny paws, he licked with pleasure, then went to the other glasses and imbibed until he grew drunk, as the partiers laughed. He paid dearly for the spree: the next morning, Maurycy found his companion lying stiff and lifeless on the floor of his cage.

Chapter 31
1944

Nothing had changed in the villa's roster or routines, but a new malaise hung in the air, Antonina thought, as everyone went about their chores with a friendly smile, while trying to hide scorched nerves. People seemed "distracted," and "conversations stumbled, sentences fell apart mid-word." On July 20, a bomb planted by Count von Stauffenberg exploded at Hitler's *Wolfschanze* (Wolf's Lair) headquarters in the Prussian forest, though Hitler escaped with only minor injuries. After that, panic grew in the local German population, and columns of retreating soldiers began streaming through Warsaw, blowing up buildings as they fled westward. Gestapo members burned their files, purged warehouses, and shipped personal belongings back to Germany. The German governor, mayor, and other administrators bolted away in any handy truck or cart, leaving only a garrison of 2,000 soldiers behind. As the Germans rushed out, creating a void, many Poles hurried in from nearby villages, afraid the coming soldiers might ransack their houses or farms.

Convinced the Uprising would start any minute, Jan felt sure it would cost only a few days at most for the 350,000 men of the Home Army to overwhelm the remaining Germans.

In theory, once the bridges were captured by the Poles, battalions from both sides of the Vistula River could join ranks and create one single powerful army to liberate the city.

On July 27, when Russian troops reached the Vistula sixty-five miles south of Warsaw (Antonina said she could hear the gunfire), German Governor Hans Frank summoned 100,000 Polish men between the ages of seventeen and sixty-five to work nine hours a day building fortifications around the city, or be shot. The Home Army urged everyone to ignore Frank's order and start preparing for battle, a call to arms echoed the next day by the Russians, pushing closer, who broadcast in Polish: "The hour for action has arrived!" By August 3, as the Red Army bivouacked ten miles from the right-bank district that included the zoo, life grew even tenser in the villa and people kept asking: "When will the Uprising start?"

The dramatis personae at the zoo changed abruptly. Most Guests had already left to join the army or escaped to safer *melinas*: Fox Man planned to move to a farm near Grójec; Maurycy joined Magdalena in Saska Kępa; and although the lawyer and his wife fled to the other side of Warsaw, their two daughters, Nunia and Ewa, decided to stay at the villa, because if something were to happen to Antonina, they insisted, then newborn Teresa, Ryś, Jan's seventy-year-old mother, and the housekeeper would have to manage all alone, which wasn't feasible. Although soldiers started evacuating civilians from the lands closest to the river, Jan hoped his family could remain in the zoo, since Polish soldiers were bound to win the Uprising soon, and the strain of traveling might kill the baby or Jan's infirm mother. In his testimony to the Jewish Institute, he recalled that at 7 A.M. on August 1, a girl came to summon him for battle. This would have been someone like the Home Army messenger Halina Dobrowolska (during the war, Halina Korabiowska), whom I met one sunny summer afternoon in Warsaw. Now a lively woman in her eighties, she was a teenager during the war, and she remembers the day she was dispatched by bicycle and tram on a long, dangerous journey into the suburbs to summon fighters and warn families that the Uprising was due to start. She needed to take a trolley and finally found one, though the conductor was packing up, since most Warsawians had already abandoned their jobs and raced home to prepare for battle. Anticipating just such a problem, the Underground had supplied Halina with American dollars, which the conductor accepted, and he nervously drove her to her destination.

Jan raced upstairs to where Antonina slept with Teresa, and told her the news.

"Yesterday, you had different information!" she said anxiously.

"I don't understand what's happening either, but I have to go and find out."

Their friend Stefan Korboński, who was also surprised by the timing of the Uprising and not given warning, captures some of the fervor and haste on the downtown streets that day:

The tram-cars were crowded with young boys. . . . On the sidewalks, women in twos and threes were walking along briskly, with obvious haste, carrying heavy bags and bundles. "They are transporting arms to the assembly points," I muttered to myself. A stream of bicycles flowed along the roadway. Boys in top boots and wind-jackets were pedalling as hard as their legs could go. . . . Here and there was a German in uniform, or a German patrol, proceeding on its way without seeing anything, and without knowing what was happening around it. . . . I passed numerous men scurrying, grave and purposeful, in all directions, and exchanging glances with me full of tacit understanding.

Four hours later, Jan returned home to say goodbye to Antonina and his mother, explaining that the Uprising would be starting any moment. He handed Antonina a metal mess tin and said:

"There's a loaded revolver in here, just in case German soldiers show up. . . ."

Antonina froze. "I was paralyzed in place," she wrote, and said to Jan: "German soldiers? What are you thinking? Did you forget what we believed only a few days ago, that the Underground army was supposed to win? . . . You don't believe anymore!"

Jan replied grimly: "Look, a week ago we had a good chance of winning this battle. It's too late now. The timing isn't right for the Uprising. We should wait. Twenty-four hours ago, our leaders thought the same. But last night they suddenly changed their minds. This kind of indecisiveness can lead to very bad consequences."

Jan didn't know that the Russians, supposed allies, had their own voracious agenda, and that Stalin, who had been promised a chunk of Poland after the war, wanted both the Germans and the Poles to be defeated. Meanwhile, he refused to allow Allied planes headed for Poland to land on Russian airfields.

"I hugged Jan tight, my face pressed hard against his cheek," Antonina recalled. "He kissed my hair, looked at the baby, and then ran downstairs. My heart was pounding like crazy!" She hid the tin with the revolver under her bed and went to check on Jan's mother, whom she found sitting in an armchair, saying her rosary beads, "her face wet with tears."

Jan's mother would have abided by the custom of making a quick cross on her forehead and inviting Mary to bless Jan's journey. Our Lady of the Home Army (the Virgin Mary) was patron saint for the soldiers during the Uprising, when one found hastily built altars to her in the city and shrines along the roadways (Poland still has many today). Soldiers and their families also prayed to Jesus Christ, and often carried in their wallet a small portrait of Christ with the inscription *Jezu, ufam tobie* (In Jesus we trust).

We don't know what Antonina did to ease the pincers of uncertainty, but Jan once informed a journalist that she had been raised a strict Catholic, and since she'd had both children baptized, and always wore a medallion around her neck, she most likely prayed. During the war, when all hope had evaporated and only miracles remained, even unreligious people often turned to prayer. Some of the Guests used fortune-telling to help shore up morale, but as a self-proclaimed man of reason, and the son of a frankly atheist father, Jan frowned on superstition and religion, which means Antonina and Jan's devoutly Catholic mother may have kept some house secrets of their own.

As airplanes flew low strafing runs over the city, Antonina

tried to guess what was happening on the other side of the Vistula, and finally went up to the terrace, from which she searched the bright sizzle of gunfire across the river, reading every snap as a clue. The shots sounded "separate, personal," she noted, not like the streaming echoes of a big military battle.

The leadership of the zoo's little fiefdom fell to her, she realized, including Ryś, four-week-old Teresa, the girls Nunia and Ewa, her mother-in-law, the housekeeper, Fox Man and his two helpers. The "heavy ballast of being responsible for the lives of others" slid around her body and stole through her mind as obsession:

> The seriousness of the situation didn't let me relax for a moment. No matter if I wanted to or not, I had to take a leadership of our household . . . be on alert all the time like I was taught in my Girl Scout years. And I knew that Jan had much more difficult duties. I had a powerful feeling of being responsible for taking care of everything at home; I carried those thoughts obsessively. . . . I just knew I had to do it.

Sleep surrendered to war, and for twenty-three nights she forced herself to stay awake, terrified that she might doze off and not hear a noise, however tiny, that signaled danger. In some ways, this guardian spirit wasn't new to Antonina, who remembered how, during the shellings of 1939, she had shielded her young son with her body. It sprang from the ferocity of motherhood, she decided, the instinct to battle if need be to protect one's family.

Even though the battlefield lay across the river, she smelled death, sulfur, rot on westerly breezes, and heard the relentless clash of guns, artillery shells, and bombs. Without news or contact with the rest of the city, Antonina imagined the villa transformed "from an ark to a tiny ship on a vast ocean, hopelessly adrift without compass or rudder," and she expected a bomb at any moment.

Stationed on the terrace, she and Ryś craned to see the fires across the river and divine events. At night, they watched bright sparks of gunfire—single shots, not the rapid echoes of a field battle—and airplanes whining and whistling above the city until early morning.

"Dad is fighting in the worst part of the city," Ryś kept repeating as he pointed toward the Old City. For hours he stood sentry's watch, scouting the battle through binoculars, searching for his father's shape, ducking down whenever he heard a bomb growling toward him.

Just outside the door to Antonina's bedroom, a metal train ladder led up to the flat roof, and Ryś often climbed it, binoculars in hand. Germans stationed in Praski Park had taken over a small amusement park near the bridge that included a tower for parachute-jump rides, from which they spotted Ryś atop the roof, spying on them. One day, a soldier stopped by to threaten Antonina that if he ever caught Ryś up there again he'd shoot him.

Despite the skittish, sleepless nights and the daily alarms, Antonina confessed to feeling "chills of excitement" about the Uprising, "having imagined this day through the long ghastly years of occupation," though she could only guess at events. Across the river, in the heart of the city, food and water were scarce, but there was plenty of lump sugar and vodka (filched from German supplies) to fuel the Home Army as they built antitank barricades from paving stones. Of the 38,000 soldiers (4,000 were women), only one in fifteen had adequate weapons; the rest used sticks, hunting rifles, knives, and swords, hoping to capture enemy weapons.

Because the Germans still held the telephone exchange, a corps of brave girl couriers carried messages around the city, just as they had been doing secretly during occupation. When Halina Korabiowska returned to Warsaw, she headed downtown to help relay messages, set up field kitchens and hospitals, and supply the fighters.

"There were barricades everywhere," Halina told me with excitement in her voice. "Everyone was happy in the beginning. At 5 P.M. the Uprising started and we put on red-and-white armbands. . . . In the early weeks of the Uprising, we survived on one meal a day of horsemeat and soup, but by the end we ate only dried peas, dogs, cats, and birds.

"I saw my fifteen-year-old friend carrying one end of a stretcher with a wounded soldier on it. A plane flew over and she saw the fear in the soldier's eyes and lay down on top of him—she was badly wounded in the neck. Another day, on my messenger run, I encountered two women carrying heavy bags from a building. I stopped to ask if they needed help, and they said they'd found a cache of German medicines and also a huge sack of candy, some of which they offered me. I filled my jacket pockets and sleeves with candies and went among the soldiers holding my arms just high enough not to spill them. Whenever I encountered soldiers, I told them to put their hands together and I extended my arms and let candies pour into their hands!"

With the Germans in retreat, everyone could move and talk freely for the first time in years, Jews could emerge from hiding since the racist laws had evaporated, and people flew the Polish flag from their houses, sang patriotic songs, and wore red-and-white armbands. Feliks Cywiński commanded a brigade of soldiers that included Samuel Kenigswein, who led a battalion of his own. Warsaw's long-suppressed cultural life started to bloom again, cinemas reopened, literary periodicals suddenly reappeared, concerts sparkled in elegantly furnished sitting rooms. A free postal service issued stamps—Boy Scouts ran it and hand-delivered letters. An archival photograph shows a metal mailbox decorated with both an eagle and a lily, to signify that the youngest scouts risked their lives delivering its letters.

When news of the Uprising reached Hitler, he ordered Himmler to send in his harshest troops, kill every Pole, and pulverize the whole city block by block, bomb, torch, and bulldoze it beyond repair as a warning to the rest of occupied Europe. For the job, Himmler chose the most savage units in the SS, composed of criminals, policemen, and former prisoners of war. On the Uprising's fifth day, which came to be known as "Black Saturday," Himmler's battle-hardened SS and Wehrmacht soldiers stormed in, slaughtering 30,000 men, women, and children. The following day, while packs of Stukas dive-bombed the city—in archival films, one hears them whining like megaton mosquitoes—ill-equipped and mainly untrained Poles fought fiercely, radioed London to air-drop food and supplies, and begged the Russians to launch an immediate attack.

Antonina wrote in her diary that two SS men opened the door, guns drawn, yelling: *"Alles rrrraus!!"*

Terrified, she and the others left the house and waited in the garden, not knowing what to expect but fearing the worst.

"Hands high," they yelled. Antonina noticed the men's first fingers cupping triggers.

Holding her baby in her arms, she could only raise one hand, and her brain had trouble "registering their vulgar, brutal sentences" as they bellowed:

"You'll pay for the deaths of our heroic German soldiers being slaughtered by your husbands and sons. Your children"— they pointed to Ryś and Teresa—"suck in hatred for the German people along with their mother's milk. Up till now we let you behave that way, but enough is enough! From now on, one thousand Poles will be killed for every dead German."

"Surely this is the end," she thought. Hugging her baby tight, mind darting to think up a plan, she felt her heart caged in her ribs, and her legs became too heavy to move. It wasn't the first time she literally froze in fear. On this occasion, although she couldn't move, she knew she had to say something, anything, and stay calm, talk to them the way she used to soothe angry animals and gain their trust. Her mouth filled with German words she didn't think she knew, and she began talking about ancient tribes and the grandeur of German culture. As she hugged the baby tighter, words streamed from her mouth, and, in another chamber of mind, she concentrated hard and repeated over and over the command: *Calm down! Put the guns down! Calm down! Put the guns down! Calm down! Put the guns down!*

The Germans continued yelling, which she didn't hear, and they never lowered their guns, but in a spill of cobbled thoughts she kept talking while issuing silent commands.

Suddenly a soldier looked at Fox Man's fifteen-year-old helper and barked at him to go behind the shed in the garden. The boy started walking, followed by an SS man who reached into his pocket, pulling out a revolver as they disappeared from view. A single gunshot.

The other German told Ryś: "You're next!"

Antonina saw her son's face shriek with fear, the blood drain out of it, and his lips turn a light purple. She couldn't move and risk their killing her and Teresa, too. Ryś raised his hands and started to walk slowly, robotically, "as if life had already left his little body," she remembered later. Watching until he disappeared from view, she continued following him in her mind's eye: "Now he'll be close to the hollyhocks," she thought, "now he'll be near the study window." A second shot. It felt like "a bayonet plunged into my heart . . . and we heard the third shot . . . I couldn't see anything; my vision became blank, then dark. I felt so weak, I was close to fainting."

"You sit down on the bench," one German told her. "It's difficult to stand up with a child in your arms." A moment later the same man called:

"Hey, boys! Bring me that rooster! Get him from the bushes!"

Both boys ran out from the shrubs, shaking with fear. Ryś was holding his dead chicken, Kuba, by its wing, and Antonina stared fixated at big drops of blood dripping from Kuba's bullet wounds.

"We've played such a funny joke!" a soldier said. Antonina saw their marble faces loosen into laughter as they left the garden, carrying the dead chicken, and she watched Ryś slink low while trying hard not to cry, until it was no use and tears flooded him. What could a mother do to comfort a child after that?

I walked over to him and whispered in his ear: "You are my hero, you were so brave, my son. Would you please help me go inside now, because I'm very weak." Maybe the responsibility would help defuse some of his emotions. I knew how hard it was for him to show his feelings. Anyway, I needed him to steady me and the baby, since my legs really had softened from shock.

Later, when she calmed down, she tried to diagnose the behavior of the SS soldiers—did they ever consider shooting them, or was it always a sick game of power and fear? Certainly they hadn't known about Kuba, so they must have been improvising as they went along. She couldn't fathom their sudden tenderheartedness in urging her to sit down. Were they really worried that she might collapse holding her newborn? "If so," she thought, "maybe their monstrous hearts contain *some* human feeling; and if that's so, then pure evil doesn't really exist."

She'd been so sure the gunshots had killed the boys, that Ryś lay crumpled on the ground with a bullet in his head. A mother's nervous system derails at such a time, and even though they'd all survived, she found herself sinking into a savage depression, which she berated herself for in her diary: "My weakness shamed me," at precisely the time "I needed to be a leader of my little group."

In the days that followed, she also suffered headaches from

the infernal racket of the German army amassing rows of rocket launchers, mortars, and heavy artillery near the zoo. The seismic ructions of bombs followed, with shells of every caliber and shape delivering their own fiendish din: whistling, blasting, crackling, banging, crashing, scraping, thundering. Then there were the *screaming meemies*, army slang (inspired by French girls named Mimi) for a type of German shell that made a shrieking noise in flight, a term extended, in time, to battle fatigue caused by long exposure to enemy fire.

The Germans also shot mine-throwers known as "bellowing cows," which yowled six times in a row as six mines cranked into position before a series of six explosions.

"I will never forget that sound to the end of my days," wrote Jacek Fedorowicz, who was seven years old during the Warsaw Uprising. "After the cranking there was nothing one could do. If one heard the explosion, it meant one had not been killed.... I had a good ear for discerning the death-dealing sounds." He managed to escape with the "remnants of my family's fortune . . . sewn inside [my teddy bear] in the form of 'piglets,' or gold five-rouble coins. Apart from him, the only things I managed to salvage after the Uprising were a drinking glass and a copy of *Dr. Dolittle*."

Airplanes bombed the fighters in Old Town; soldiers machine-gunned Polish civilians in the streets; demolition crews torched and bombed huge buildings. The air filled with dust, fire, and sulfur. When it grew dark, Antonina heard an even scarier rumble from the direction of Kierbedź Bridge, the growl of a giant machine. Some people said the Germans had built a crematorium to burn the bodies of the dead, to protect Warsaw from plague, while others thought the

Germans had unleashed a huge radioactive weapon. The river water reflected a pale green fluorescent light so brilliant she could see people standing at their windows on the other side of the river, and after sunset, the otherworldly rumble was joined by invisible choruses of drunken soldiers who sang late into the night.

According to Antonina, she lay awake all that night, scared cold, aware of the tiny hairs stiffening on the back of her neck. As it turned out, the weird light was far less sophisticated than she had imagined; in Praski Park, the Germans had installed a generator that drove colossal reflector lamps to dazzle the enemy.

Even after the battle moved out of the zoo district, soldiers invaded the zoo to prowl and pillage. One day a gang of Russians arrived with "wild eyes," and busily began searching the cupboards, walls, and floors for anything they could steal, including picture frames and carpets. When she approached them and silently stood her ground, she sensed scavengers darting around her "like hyenas" racing into the rooms. "If they guess my fear, they'll devour me," she thought. Their leader, a man with Asian features and icy eyes, walked up close and stared hard at her, while Teresa slept nearby in a tiny wicker cradle. Antonina resolved not to look away or move. Suddenly, he grabbed the small gold medallion she always wore around her neck, "and flashed his white teeth." Slowly, gently, she pointed to the baby, then, defrosting the Russian of her childhood, commanded in a loud, stern voice:

"Not allowed! Your mother! Your wife! Your sister! Do you understand?"

When she placed her hand on his shoulder, he looked

surprised, and she saw the manic fury draining from his eyes, his mouth relaxing, as if she'd smoothed the fabric of his face with a hot iron. Her mind-whispering had worked again, she thought. Next he placed his hand into the back pocket of his pants, and for a horrible instant she remembered the German soldier with his revolver aimed at Ryś. Instead, he withdrew his hand and opened it to reveal several dirty pink hard candies.

"For the baby!" he said, pointing to the cradle.

As Antonina shook his hand in thanks, he smiled at her admiringly and glanced at her ringless hands, then made a pitiful face, took a ring off his own finger, and offered it to her.

"It's for you," he said. "Take it! Put it on your finger!"

Her heart "shook" as she slipped on the ring, because it bore a silver eagle, a Polish emblem, which meant he'd probably ripped it off the finger of a dead Polish soldier. "Whose ring was it?" she wondered.

Then, loudly summoning his soldiers, he ordered: "Leave everything you took! I will kill you like dogs if you don't obey me!"

Surprised, his men dropped all the furniture and loot they'd gathered and dragged small items out of their pockets.

"Let's go now—don't touch anything!" he said.

With that, she watched his men "shrink in size as they left one at a time like muzzled dogs."

When they'd gone, she sat down at the table and looked again at the ring with the silver eagle and thought: "If felt words like *mother*, *wife*, *sister*, have the power to change a bastard's spirit and conquer his murderous instincts, maybe there's some hope for the future of humanity after all."

From time to time, other soldiers visited the zoo, without incident, and then one day a car pulled up with several German clerks who managed Third Reich fur farms and knew Fox Man from his days in Grójec. Fox Man reported that the animals still survived with luxuriant fur, and they gave him permission to move both animals and employees to Germany. Packing up so many animals would take time, which meant everyone could stay in the villa for a while longer, possibly even until the Uprising triumphed and the Germans deserted Warsaw. Then no one would have to leave the zoo at all.

In the meantime, trying to weaken the Resistance, German airplanes kept dropping notes urging Warsaw's civilians to abandon the city before it was gutted. Soon afterward, the German army trucked even more heavy artillery into Praski Park, hiding it among the trees and bushes near the river. Stationed so close, German soldiers often stopped by for a drink of water, a cup of soup, or some cooked potatoes. One evening, a tall young officer expressed concern about civilians living too close to the battlefield, and Antonina explained that she and the others were running a high-priority Wehrmacht fur farm, which they couldn't leave because it was an inauspicious time for the raccoon dogs, which develop soft dense coats by molting in summer and then regrowing a winter pelage during September, October, and November. Tamper with their schedule by boxing them up, stressing them, and shipping them to a different climate, she warned, and their prized winter coats wouldn't grow back soon. That seemed to satisfy him.

Thunder had never frightened her before, she wrote— "After all, it's only sound filling the vacuum created by streaks of lightning"—but the artillery flared without letup, the air didn't grow moist as a prelude to storm, no rain fell, and the dry thunder jangled her nerves. One afternoon, the artillery suddenly paused, and during that rare lull the women of the house lay down and rested, soaking in the quiet. Jan's mother, Nunia, and Ewa all took a nap in their bedrooms, and Antonina nursed Teresa downstairs, on a sweltering day, with all the doors and windows open. Suddenly the kitchen door squeaked and a German officer strode into the room. He stopped for a moment when he saw her with the baby, and as he edged closer Antonina smelled alcohol on his breath. Snooping around suspiciously, he wandered into Jan's study.

"Oooh! A piano—sheet music! Do you play?" he asked excitedly.

"A little," she replied.

Paging through some Bach, he paused and started to whistle a fugue with tuneful expertise. She figured him for a professional musician.

"You seem to have a perfect ear for music," she said.

When he asked her to play for him, she sat down at the piano, though something didn't feel quite right. Tempted to grab Teresa and make a run for it, she feared he'd shoot her if she tried, so instead she began playing "Ständchen," a romantic song by Schubert, hoping that this German favorite might calm him with sentimental memories.

"No, not that! Not that!" he screamed. "Why are you playing *that*?!"

Antonina's fingers sprang from the keys. Clearly a wrong choice, but why? She'd heard and played the German serenade so often. As he strode to the bookshelf to page through sheet music, she glanced down and read the lyrics to "Ständchen" :

> *Softly through the night my songs implore you.*
> *Come down into the still grove with me, beloved;*
> *Slender treetops rustle and whisper in the moonlight.*
> *Fear not, sweet one, the betrayer's malicious eavesdropping.*
> *Do you hear the nightingales calling? Ah! They are*
>   *imploring you,*
> *With the sweet music of their notes they implore you for*
>   *me.*
> *They understand the bosom's yearning, they know the pangs*
>   *of love,*
> *They can touch every tender heart with their silvery tones.*
> *Let them move your heart also; beloved, hear me!*
> *Trembling, I wait for you; come, give me bliss!*

A broken heart, that would rattle anyone, she thought. Suddenly his face lit up as he opened a collection of national anthems, through which he searched, looking eagerly for something, which he finally found.

Placing the open book on the piano, he said: "Please, play this for me."

As she started to play, the German officer sang along, pronouncing the English words in a heavy accent, and she wondered what the soldiers in Praski Park must be thinking as he belted out "The Star-Spangled Banner." Occasionally she peeked up at his half-closed eyes. When she finished with a flourish, he saluted her and quietly left the villa.

Who was this officer so fluent in music, she wondered,

and what was the American anthem all about? "Maybe he was joking with another German sitting near the villa?" she thought. "Surely someone will come and interrogate me about the music? Now I'll have to worry about provoking the SS." Later, she decided that he probably meant to terrorize her, and, if so, it had worked, because the melody snagged in her head and kept repeating until a round of cannonades split the night.

As the Germans stepped up their attack on Old Town, Antonina still hoped the Underground army would win, but rumors of Hitler's order to demolish the city trickled in. Soon she learned that Paris had been liberated by the Free French, U.S., and British forces; and then Aachen, the first German city to fall, devastated by 10,000 tons of bombs.

She had no word from or about Jan, stationed in Old Town, where the Home Army, forced into a smaller space, fought from building to building, even room to room in a house or cathedral. Many witnesses tell of the front suddenly erupting inside a building and flowing from floor to floor, while those outside faced a continuous shower of bombs and bullets. All Antonina and Ryś could do was watch the heavy gunfire ricocheting around Old Town and picture Jan and their friends moving along cobblestone streets she knew by heart.

In an archival photograph taken by field reporter Sylwester "Kris" Braun on August 14, Polish soldiers are proudly displaying a German armored personnel carrier they have just captured. Jan is not in the photograph, but it can hardly be sheer coincidence that, as the caption notes, they nicknamed the elephantine vehicle "Jás," the same name as the Warsaw Zoo's male elephant, killed early in the war.

Chapter 32

By September, 5,000 soldiers in Old Town had escaped through the sewers, despite Germans dropping grenades and burning gasoline down the manholes. Elsewhere, the Allies were advancing on all fronts: after liberating France and Belgium, the United States and Britain were pushing into Germany from the Netherlands, Rhineland, and Alsace; and though the Red Army paused near Warsaw, it had already captured Bulgaria and Romania, was prepared to take Belgrade and Budapest, and planned to storm the Reich from the Baltics; the United States had landed on Okinawa and was pounding the South Pacific.

A German officer assured Fox Man that, whatever happened to the military, the Third Reich needed its valuable fur farms and he should prepare to pack his animals into well-vented crates and move them to a small town in the suburbs for safety. As shells started falling closer to the zoo, Antonina prepared to uproot her household, too, and the nearby town of Lowicz, where Fox Man was headed, seemed a haven out of battle range but still close to the city. Antonina, Ryś, Jan's mother, the two girls, Fox Man, and his helpers planned to travel together, hoping all would pass as fur farm workers.

Choosing which pets to leave behind tormented them (muskrat, Wicek, other rabbits, cat, dog, eagle?), but in the end they decided to risk taking only Wicek and release all the rest to the wild and their wits.

Although they could cart whatever household items they pleased, it seemed prudent to travel light, so they bundled up only mattresses, comforters, pillows, winter coats, boots, water containers, pots, shovels, and other practical items. Anything of value had to be hidden far from bombs and prowling soldiers; they loaded the fur coats, silver, typewriter, sewing machine, documents, photographs, heirlooms, and other treasures into large boxes, and Fox Man and his boys quickly stashed them in the underground corridor leading from the villa to the Pheasant House, then they bricked up the entrance to the tunnel.

On August 23, the day of departure, Ryś watched as a huge shell landed about fifty yards from the villa, and dug in but didn't explode; a bomb squad appeared soon afterward with an officer who swore that anyone still in the villa at noon would be shot. Ryś ran to the Pheasant House and fed the rabbits dandelion leaves a last time, then opened all the cages and turned them loose. Confused by their newfound freedom, the rabbits refused to leave, so Ryś lifted them out by their long ears, one at a time, and carried them to the lawn. No predators lurked in the brush, ponds, or sky, and the last of the household pets—eagle and muskrat—had been freed the day before.

"Go, silly rabbits, go!" Ryś said, shooing them. "You're free!"

Antonina watched fur balls of all sizes hopping slowly through the grass. Suddenly Balbina sprang from the bushes and ran to Ryś with a high-flagging tail and a loud purr. One whiff of the cat and the rabbits bolted, as Ryś lifted Balbina into his arms.

"What! Balbina, do you want to go with us?" Carrying her, he walked toward the house, but the cat squirmed free.

"You *don't* want to go with us? Too bad," he said, adding bitterly, "But you're lucky, at least *you* can stay here." She sloped away between the bushes.

Watching this scene from the porch, Antonina felt a powerful desire to stay home, too, accompanied by an equally strong wish for the truck to arrive that would carry them to the train station, checking her watch over and over, though "the watch hands moved without pity." The impulse to plunge into some bolt-hole in Warsaw flashed through her mind, but where would they go? She worried about her lame mother-in-law, "who couldn't walk half a mile," and being waylaid by Germans, who, she'd heard, were arresting every Pole they could find and shipping all to a death camp near Pruszków. As things stood, traveling west with the fur farm animals made the most sense.

At last, at 11:30 A.M., Fox Man's old truck clattered up to the villa, and they quickly stowed their luggage. Leaving the zoo behind, they wove through back streets until they reached the train station where a freight car waited, already loaded with foxes, minks, nutrias, raccoon dogs, and Wicek. Antonina and the others boarded, and soon the train crossed the river, paused at a couple of stations to pick up more passengers, and finally set off slowly. In Lowicz, they were told to unload their crates and await the arrival of fur animals from elsewhere in Poland, and then the assembled stock would travel to one large farm in Germany. Antonina spent the day strolling through the village, struck by her liberty and the momentous quiet of a town showing no signs of war. The next day she went looking for help and learned that Andrzej Grabski, son of the Polish ex-prime minister, happened to be on the German fur company board; when she explained that she feared taking her small children to Germany, Grabski found a temporary shelter for her in town. Six days later she said goodbye to Fox Man (who had to remain in Łowicz with the animals), rented a horse-drawn wagon, and headed for the village of Marywil, only four miles away, yet "a long, slow journey that felt like forever."

When they finally arrived at a little schoolhouse on an old estate, a woman offered them a small classroom to sleep in, whose wooden walls were splotched with dirt and floor strewn with mud and straw. Cobwebs hung from the ceiling, all the windowpanes were broken, and piles of cigarette butts littered the floor. They set Wicek's cage beside a clay stove, and Antonina wrote that his scratching to get out created the only noise in a vault of silence that seemed bizarre after weeks of explosions and gunfire, not a calming silence but vacant, unnatural, disturbing, "a nuisance to our ears."

"The quiet is spooky," Ryś said, wrapping his arms around her neck and hugging her tight. Although she didn't want him scared or suffering, she wrote that it felt wonderful to have him need her soothing. During the uncertain and violent days of August, she'd watched him trying to act strong and grown-up, but now, to her relief, "at last he could let himself be a child."

"Mom, I know we're never going home again," he said tearfully.

Moving from a large old city at war to a peaceful hamlet where there was no point settling in for what they anticipated as a short stay, they'd lost contact with friends, family, and the Underground, but they also lost the frights of artillery. Haunted by a distant underpinning of her world, Antonina described feeling "whipped by a disaster [I] couldn't name or influence . . . unreal and floating" much of this time, though she vowed to buoy up Ryś's spirits.

On a hunt for broom, rags, and bucket, they knocked on the door of a room where Mrs. Kokot, the local teacher, her blacksmith husband, and their two boys lived. A short, solid woman with dimples and work-worn hands greeted them.

"I'm sorry," Mrs. Kokot said, "that we didn't have time to clean the classroom before you moved in. My husband will stop by tomorrow and install a proper stove. Don't worry, everything will be all right. You'll settle in soon and feel at home here."

Over the next few days, Mrs. Kokot provided bread and butter, and brought a small wooden bathtub for Teresa and hot water. Soon life didn't seem quite as dire, but Antonina worried about Ryś, who had "lost everything he knew . . . like a tiny piece of grass uprooted by a strong wind and blown far away from its garden." What with "the earthquake of leaving Warsaw, worries about his father," whom they'd had no news of, "and all the unknowns, and the poverty," it didn't surprise her when he became depressed and moody.

But as days passed, Ryś grew closer to the Kokot family, whose daily routines yielded order and a predictability he craved. Antonina worried that, having acted more adult than child for most of the war, Ryś had gotten to the point that "he flatly refused to accept childhood, and whoever treated him like a child drew a rude response." But the mundane events of Kokot family life, in which children went to school and played without fear, proved a tonic. As he watched them going about their lives, she noted, he admired how well they worked together as a family, and also performed many charitable acts—Mrs. Kokot would ride her bike to the village to give a sick person an injection, or as far as the city to bring a doctor; and her husband would fix neighbors' engines, sewing machines, rubber wheels, watches, lamps, or any other *sick* objects.

"Ryś never made much of intellectuals," Antonina mused, "being absorbed in abstract ideas seemed silly to him. He admired practical know-how, and so he deeply respected the Kokots for their talents, common sense, and hard work." Shadowing Mr. Kokot all day, he helped replace panes of broken glass, filled cracks in wooden window frames with moss and straw, and plugged the holes in walls by using straw caulking or a mixture of lamp oil and sand.

Then Ryś did something surprising. As the ultimate sign of friendship, he gave his beloved rabbit Wicek to the Kokots' sons, Jędrek and Zbyszek. This extraordinary gesture didn't alter Wicek's living conditions much, since the boys played together all the time, but the privilege of feeding Wicek and piloting his future changed hands. At first, Antonina noted, Wicek didn't understand what was happening. Then she heard Ryś giving him a serious and detailed explanation of who his new owners were and where he'd be sleeping; afterward Wicek kept trying to steal back into Ryś's room, only to be turned away at the door.

"Now you live in Jędrek and Zbyszek's apartment, you silly creature!" Ryś said. "Why don't you want to understand this simple thing?"

Antonina watched the rabbit listening to Ryś, moving his ears and looking at Ryś "as if he understood perfectly well," but the minute Ryś carried him into the hallway between the apartments, set him down, and closed the door behind him, Wicek began scratching at the door to return.

Depression scathed Antonina once more, which she matter-of-factly recorded, without fuss or details, as if it were just another form of weather. The trip had been so depleting that, "like someone in a trance," she pushed herself to secure food and help for her small tribe of women and children. Somehow she finagled potatoes, sugar, flour, and wheat from a woman in the village; peat to use as fuel from a man down the road; and half a liter of milk a day from the county.

The spirited Warsaw Uprising collapsed after sixty-three days of ferocious street-to-street fighting, with much of the city in rubble, when what was left of Warsaw's Home Army surrendered, in exchange for the promise of humane treatment as prisoners of war, not partisans. (Nonetheless, most survivors were shipped off to slave labor camps.) Overflowing hospitals were burned with patients still in them, and women and children were roped onto tanks to prevent ambush from snipers. Hitler celebrated by ordering Germany's churches to ring their bells for a week solid.

The roads streamed with refugees seeking shelter in the neighborhood of Łowicz and Marywil, a countryside dotted with feudal estates, complete with manor houses, small poor farms, hamlets the landowners helped support, and many

Diane Ackerman

locals employed at the manors. Day by day, more people swarmed into the region, until farmers, overwhelmed by the sheer mass of hungry, frightened people landing in their fields and on their doorsteps, begged local officials to relocate them elsewhere.

When Antonina and her family had first arrived at the schoolhouse, they tried to lie low, in case the Gestapo might be chasing them, but as days passed quietly, they began to relax, and after a few weeks in Marywil, following Warsaw's capitulation, they started angling for news of family and friends. Antonina awaited word of Jan, convinced he'd magically appear one day, "having moved heaven and earth" to find her, as he had with Dr. Müller's help in 1939. She knew nothing of Jan's bizarre luck during the early days of the Uprising, when he was shot through the neck and rushed to the hospital on Chmielna Street, to die, everyone thought, since it's nearly impossible for a bullet to fly through someone's neck without hitting the esophagus, spine, veins, or arteries. Years later, Antonina met the doctor who had treated him. "If I had anesthetized him," Dr. Kenig recalled in amazement, "and *tried* to re-create the route of that bullet, I couldn't do it!" When Germans captured the hospital, he was shipped to a POW camp for officers, where he mended from the bullet wound only to battle hunger and exhaustion.

Antonina sent a letter to a family friend, who agreed to relay messages for her; and Nunia, who, instead of joining her own parents, had stayed with Antonina and Ryś to help look after things and act as messenger, rose before dawn one morning, waited hours for the horse and wagon that served as a "bus," and traveled to Warsaw by way of Łowicz. All along the route, she tacked up small pieces of paper asking about Jan Żabiński and giving Antonina's address; she pinned them to trees, electric poles, fences, buildings, train station walls, in what had become a public lost and found bureau. Stefan Korboński remembers how

> on the fences of all the stations were hundreds of notices and the addresses of husbands searching for their wives, parents searching for their children, and people in general announcing where they were. Large crowds stood in front of these "forwarding offices," from morning till night.

Soon Antonina started receiving letters with clues: from the nurse at the hospital where Jan was treated for his neck wound, a mailman in Warecki Square, a guard at the Zoological Museum on Wilcza Street. All wrote to tell of Jan and give her hope, and when she learned he'd been shipped to a German POW camp, she and Nunia wrote dozens of letters to all the camps that imprisoned officers, fishing for leads.

Chapter 33
December, *1944*

With winter, the endless mud puddles froze over and the land grew firm and fibrous again under a slather of white as Antonina prepared a Christmas starkly unlike those before the war. On Christmas Eve, Poles traditionally served a twelve-dish meatless dinner before exchanging gifts, and the zoo's Christmas Eve used to include a special bounty. Antonina remembered how "a wagon drove into the zoo full of unsold Christmas trees; it was a gift for the zoo animals: ravens, bears, foxes, and many other animals liked to chew or peck at the aromatic bark or needles of evergreens. Christmas trees went to different aviaries, cages, or animal units, and the holiday season officially started in the Warsaw Zoo."

All night, comets of lantern light would orbit the grounds: one man dutifully guarding the exotic animals section, checking the heat in buildings, and adding coal to the furnaces; several men carrying extra hay to barns and open shelters; others tucking extra straw into the aviaries where the tropical birds burrowed in to stay warm. It had been a scene of refuge and dancing lights.

This Christmas Eve, 1944, as Ryś headed for the woods with Zbyszek, he announced to Antonina that "children should have a little fun." Later the boys returned dragging two small fir trees.

Following rural custom, trees were decorated during the daylight and lit when the first planet appeared (to honor the Star of Bethlehem), then dinner was served with extra places set for absent family members. Antonina wrote of arranging the small tree on a stool, where baby Teresa found it a source of hand-clapping delight, which she babbled to as the family embellished shiny branches with "three small apples, a few gingerbread cookies, six candles, and several straw peacock-eye ornaments that Ryś had made."

Over the holidays, Genia surprised Antonina with a visit; risking arrest because of her Underground activities, she took the train, then walked through gusty cold for four miles, to bring money, food, and messages from friends. Antonina and Ryś still had no word from Jan. One day, Mrs. Kokot biked to the post office as usual, and they watched her returning, as usual: a tiny figurine growing larger and more defined as she pedaled nearer. This time she was waving a letter. Ryś ran out to meet her in his shirtsleeves, grabbed the letter, and dashed indoors with Mrs. Kokot following, smiling.

"Finally," was all she said.

After Antonina and Ryś read the letter several times, Ryś rushed away to share the news with Mr. Kokot; according to Antonina, Ryś rarely spoke of his phantom father, whom he could now risk mentioning at last.

In the modern Warsaw Zoo's archive, along with photographs donated by the family, there's a wonderful oddity: a card Jan sent to them from the POW camp, with no writing on it except the address. On the back, a good caricature of

Jan wears a baggy uniform with two stars on each epaulet, and a dark scarf knotted around his neck and flowing down past his waist. He's captured himself with stubbly beard, pouchy eyes and long lashes, heavily wrinkled brow, three wisps of hair poking up from his bald crown, a cigarette stub dangling from his mouth, and a look of boredom and disdain on his face. Nothing written, nothing incriminating, just a drawing that exists somewhere between pathos and humor, which depicts him as whipped but not defeated.

The Red Army finally entered Warsaw on January 17, long after the city's surrender and too late to help. In theory, the Russians were supposed to drive out the Germans, but for political, strategic, and practical reasons (among them, losing 123,000 men en route), they had camped on the east side of the Vistula River and complacently watched the bloodshed for two months solid, as thousands of Poles were massacred, thousands more sent to camps, and the city extinguished.

Halina and her first cousin, Irena Nawrocka (an Olympic fencing champion who had traveled widely before the war), and three other girl messengers were arrested by the Germans and ordered to march with a large bedraggled herd of guards and captives from Warsaw to a labor camp in Ożarów. Darting in from the fields, farm workers handed the girls work clothes to slip on and tools to carry, then pulled them from the crowd, between the rows of flax, before the exhausted guards noticed. Blending in with the field hands, the girls escaped to Zakopane (in the Tatra Mountains), where they hid for several months until the war ended.

Chapter 34

1945

Flocks of crows circled the sky before landing in the snow-covered fields, on one of those claggy, warm January mornings when dark tree branches glisten through fog and just breathing feels like inhaling cotton. The morning bristled with signs. Antonina heard the rumble of heavily armed trucks, the grinding of airplanes and distant explosions, then people shouting: "The Germans are running away!" Soon the Polish and Soviet armies appeared, walking together, and as a long caravan of Soviet tanks crawled by, locals quickly hoisted red flags to welcome the liberators. Suddenly a huge flock of white pigeons flew up the sky and soared above the soldiers, reassembled as a single cloud and swerved even higher. "The timing was perfect," Antonina wrote. "Surely some movie director arranged this symbolic scene."

Although she nourished hope of Jan's release, she decided to pass the rest of the winter in Marywil, because traveling to Warsaw with small children seemed risky. However, local children itched to return to school, their own private time-keeping, which meant Antonina's group had to leave the schoolhouse for another temporary shelter. When her food money ran out and she needed to buy milk for the baby, the manor house took pity on her and sent provisions. Fortunately, she had saved a few gold "piglets" (rubles) for buying their passage back to Warsaw, a trip she knew might be costly. Once again, refugees clogged the roads, this time desperate for home, even though they'd heard their apartments lay in ruins. Nunia hurried on ahead to scout, and brought back news that she'd found friends still living in the zoo district, with whom they could stay, and she reported that the villa, though blasted and looted, still stood.

Needing a large truck, a scarce commodity, Antonina prevailed upon soldiers traveling east with a load of potatoes, who agreed to carry her group part of the way. On travel day it was zero degrees, and only the baby, swaddled in a small down blanket, didn't shiver as the truck shambled along, pausing frequently to be searched by soldiers on patrol. Dropped off in Włochy, they secured a ride with a Russian pilot, who agreed to share his open truck, into which they piled.

As they finally entered Warsaw's city limits, a wave of filthy snow and sand splashed the sides of the truck, the snow stank, the sand irritated their eyes, and they huddled to keep warm. What she saw "dazed and sickened" her, she wrote, because, despite rumors, warnings, and eyewitness reports, she still wasn't prepared for a city in tatters. Archival photographs and films show charred window and door frames standing like sky portals, tall office buildings reduced to a hive of open cells, apartment houses and churches calving like glaciers, all the trees felled, the parks heaped with rubble, and surreal streets lined with façades thin as tombstones. In some shots, a sickly pale winter sun oozes into the crevices of pockmarked buildings, over raw metal cables, weirdly twisted pipes and iron. With 85 percent of the buildings destroyed, the once-ornate city looked like a colossal refuse heap and cemetery, everything rendered down to its constituent molecules, all the palaces, squares, museums, neighborhoods, and landmarks reduced to classless chunks of debris. Captions read: "dead city," "a wilderness of ruins," "mountains of rubble." Cold as the day was, Antonina wrote that she began sweating, and that night, mired in shock and exhaustion, they stayed with Nunia's friends.

After breakfast the next morning, Antonina and Ryś hurried to the zoo, where Ryś rushed ahead, then circled back, pink-cheeked from the cold.

"Mom, our house survived!" he said excitedly. "The people who said it was destroyed lied to us! It's damaged, there are no doors or floors and all our belongings were stolen, but there's a roof and walls! Mom! And stairs!"

A layer of snow masked the ground, and most of the trees had been sheared off by shells, but some delicate black branches still loomed against the blue sky, as did the Monkey House, the villa, and the ruins of several other buildings. One of the villa's upstairs rooms had completely disappeared, and all the wooden parts on the first floor were missing—doors, closets, window frames, floors—she assumed they'd been burned for warmth during the winter. The underground corridor leading between the basement and the Pheasant House, where they'd stored valuables, had not simply caved in but dematerialized (and there are no reports of anyone unearthing it after the war). A thick pastiche of damp papers and book pages littered the floor, which they couldn't avoid walking on and crushing even more. Together they dug through the sediment, collecting scraps of dirty documents and yellowed photographs, which Antonina stowed carefully in her purse.

Despite the cold, they inspected the garden, gouged from bombs and shells, and surveyed the grounds, a scene of barricades, deep antitank ditches, pieces of iron, barbed wire, and unexploded shells. She didn't venture any farther for fear of land mines.

It looked and smelled like "the war just left this place." While she planned renovations, Ryś "tested his memory" of the villa he grew up in against the barren world he now found. Antonina checked where they had planted vegetables the year before, and in one tiny spot where the wind had blown off a lid of snow, she saw a small strawberry plant near the ground. "An omen of new life," she thought. Just then something moved in a basement window.

"A rat?" Ryś suggested.

"Too big for a rat," Antonina said.

"A cat!" Ryś yelled. "It ran into the bushes and it's watching us!"

A thin gray cat crouched warily in a corner, and Antonina wondered if people had tried to capture it for the stew pot.

"Balbina? Old cat! Dear cat! Balbina, come here!" Ryś called as he crept closer, calling her name over and over until she calmed down and seemed suddenly to remember, flying to Ryś like a fur-fledged arrow and jumping into his open arms.

"Mom, we have to take her home to Stalowa Street!" Ryś pleaded. "We can't leave her here! Please!"

As Ryś walked toward the gate, the cat fidgeted to jump down.

"It's just like last summer," Ryś sulked. "She's running away!"

"Let her go," Antonina said softly. "She must have an important reason for staying, one we don't understand."

Ryś released her and she darted into the bushes, then stopped and looked back with her scrawny, half-starved face. She meowed, which Ryś translated as: "*I'm* going back home. What about *you*?"

For Antonina there was no going back to her previous life. Gone were the gaggling geese, squawking cormorants, whimpering gulls, the peacocks fanning iridescent tails as they strolled in sunlight, the Jericho-wall-tumbling groans of the lions and tigers, the trilling monkeys swinging on rope vines, the polar bears soaking in their pool, the blooming roses and jasmine, and the two "nice little happy otters which became best friends with our lynxes—instead of sleeping in their own basket . . . they napped in the soft fur of the lynxes, sucking on their ears." Gone were the days when the lynx cubs, otters, and puppies all lived indoors and played endless chase games together in the garden. She and Ryś staged a private ritual—they formally promised all the broken and abandoned objects that they "would remember them and return soon to help."

Chapter 35
Aftermath

While still in hiding, Magdalena Gross married Maurycy Fraenkel (Paweł Zieliński), and after the Warsaw Uprising they moved to the eastern city of Lublin, where artists and intellectuals gathered in the Café Paleta. There she met the city's avant-garde art world, which included many theaters without words: music theater, dance theater, drawing theater, shadows theater, and theaters featuring paper costumes, rags, or small fires. Poland's long tradition of subversive political puppet theater had dissolved during the war, but in Lublin she joined enthusiasts who dreamt up the first puppet theater for the new Poland, and they invited her to create the puppets' heads. Instead of crafting them with the traditional bold papier-mâché features, she decided to create lifelike facial nuances and adorn the puppets in silk, pearls, and beads. The first performance took place in Lublin on December 14, 1944.

In March of 1945, Magdalena and Maurycy returned to the newly liberated Warsaw, without electricity, gas, or transportation, whose few surviving houses tilted, windowless. Longing to sculpt animals again, she asked Antonina plaintively: "When will you have animals? I have to sculpt! I've wasted so much time!" Absent the flamingos, marabous, and other exotics she preferred, she began by sculpting the only available model, a duckling, and since she was a slow artist, she had to keep revising the piece as the duckling vamped into an adult bird. Still, it was her first sculpture after the war, cause for celebration.

The Warsaw they knew before the war had contained one and a half million people; in early spring of 1946, another visitor, Dr. Joseph Tenenbaum, reported "half a million at most. As it was, I could not see living space for a tenth that number. Many still lived in crypts, caverns, cellars, and subterranean shelters," but he was greatly impressed by their morale:

Nowhere in the world are people so generally reckless of danger as in Warsaw. There is incredible vitality in Warsaw and an infectious spirit of daring. The pulse of life beats in an unbelievably rapid tempo. People may be shabbily dressed, their faces worn and visibly undernourished, but they are not dispirited. Life is tense, yet undismayed and even gay. People jostle and bustle, sing and laugh with a mien of amazing cocksureness. . . .

There is a rhythm and romanticism in everything, and a bumptiousness that takes the breath away. . . . The city is like a beehive. The entire city works, tearing down ruins and building new houses, destroying and creating, clearing away and filling in. Warsaw started to dig out from the ruins the very moment the last Nazi trooper left its suburbs. It has been at it ever since, building, remodeling and restoring without waiting for plans, money or materials.

Throughout the city, he heard an aria by A. Harris, the unofficial "Song of Warsaw," whistled, sung, and blared through loudspeakers in the central squares as people worked. Its lover's lyrics pledged: "Warsaw, my beloved, you are the object of my dreams and yearning. . . . I know you are not what you used to be . . . that you have lived through bloody days . . . but I shall rebuild you to your greatness again."

Jan returned from the internment camp in the spring of 1946, and in 1947 he began cleaning and repairs, and erecting new buildings and enclosures for a revived zoo, one holding only three hundred animals, all native species donated by people in Warsaw. Some of the zoo's lost animals were found, even Badger, who had tunneled out of his cage during the bombardment and swum across the Vistula (Polish soldiers returned him in a large pickle barrel). Magdalena sculpted *Rooster*, *Rabbit I*, and *Rabbit II*, slowing down then in poor health ("damaged by the war," Antonina reckoned), and dying on June 17, 1948, the same day she finished *Rabbit II*. Her dream had always been to create large sculptures for the zoo, and Antonina and Jan wished she'd had that chance, especially since the zoo offered an ideal backdrop for large artworks. At today's zoo, the main gates greet visitors with a life-size zebra, wearing iron bars as striped bulging ribs. Some of Magdalena's sculptures now grace the zoo director's office, as well as the Warsaw Museum of Fine Arts, just as Antonina and Jan had wished.

One day before the July 21, 1949, reopening of the Warsaw Zoo, Jan and Antonina placed Gross's sculptures *Duck* and *Rooster* near the stairs to a large fountain visitors were sure to pass. July 21 fell on a Thursday that year, and they may have wished to avoid opening on Friday the twenty-second because people still associated that unlucky date with the 1942 liquidation of the Warsaw Ghetto.

Two years later Jan suddenly retired from zookeeping, though he was only fifty-four years old. Postwar Warsaw, under Soviet rule, didn't favor people who fought with the Underground, and, at odds with government officials, he may have felt obliged to retire. Norman Davies captures the mood of that time:

> Anyone who dared to praise pre-war independence, or to revere those who fought during the [Up]rising to recover it, was judged to be talking dangerous, seditious nonsense. Even in private, people talked with caution. Police informers were everywhere. Children were taught in Soviet-style schools where denouncing their friends and parents was pronounced an admirable thing to do.

Still needing to support his family, and devoted to zoology, Jan focused on his writing, producing fifty books that illuminated the lives of animals and sued for conservation; he also broadcast a popular radio program on the same topics; and he continued his efforts with the International Society for the Preservation of European Bison, which prized its small herd of bison in Białowieża Forest.

Oddly enough, those animals survived thanks in part to the efforts of Lutz Heck, who, during the war, shipped back many of the thirty bison he had stolen for Germany, along with back-bred, look-alike aurochsen and tarpans, to release in Białowieża, the idyll where he pictured Hitler's inner circle hunting after the war. When the Allies later bombed Germany, mother herds of the animals died, leaving those in Białowieża as their species' best hope.

In 1946, at the first postwar meeting of the International Association of Zoo Directors, in Rotterdam, reactivating the European Bison Stud Book fell to Jan, who began scouting the pedigrees of all bison that survived the war, including those in Germany's breeding experiments. His research documented prewar, wartime, and postwar bloodlines, and returned the program and pedigree watch to the Poles.

While Jan wrote for adults, Antonina penned children's books, raised her two children, and stayed in touch with the extended family of the Guests, who had traveled to different lands. Among those Jan personally led from the Ghetto (through the Labor Bureau building) were Kazio and Ludwinia Kramsztyk (cousins of renowned painter Roman Kramsztyk), Dr. Hirszfeld (specialist in infectious diseases), and Dr. Roza Anzelówna and her mother, who stayed in the villa for a short time, then moved to a boardinghouse on Widok Street recommended by friends of the Żabińskis. But after a few months they were arrested by the Gestapo and killed, the only Guests of the villa who didn't outlast the war.

The Kenigsweins survived the occupation and retrieved their youngest son from the orphanage, but in 1946 Samuel died of a heart attack, after which Regina and the children immigrated to Israel, where she remarried and worked on a kibbutz. She never forgot her time at the zoo. "The Żabińskis' home was Noah's Ark," Regina told an Israeli newspaper twenty years after the war, "with so many people and animals hidden there." Rachela "Aniela" Auerbach also moved to Israel, after first traveling to London, where she delivered Jan's report about the survival of the European bison to Julian Huxley (prewar director of the London Zoo). Irena Mayzel resettled in Israel, and hosted the Żabińskis there after the war. Genia Sylkes moved to London, too, then to New York City, where she worked for many years in the Yiddish Scientific Institute library.

Captured by the Gestapo and brutally tortured, Irena Sendler (who winkled children out of the Ghetto) escaped, thanks to friends in the Underground, and spent the rest of the war in hiding. Despite her broken legs and feet, she worked in Poland as a social worker and advocate for the handicapped. During the war, Wanda Englert would move many times; her husband, Adam, was arrested in 1943 and imprisoned in Pawiak Prison, Auschwitz, and Buchenwald. Amazingly, he survived prison and the concentration camps, later reunited with his wife, and together they moved to London.

Halina and Irena, the girl messengers, still live in Warsaw today and keep in close touch, best friends for over eighty-two years. On the wall in Irena's apartment, along with her fencing medals, are photographs of her and Halina as young women, in which they're coiffed, glamorous, and all future—studio portraits taken during the war by a neighbor.

Sitting with Halina in the courtyard restaurant of the Bristol Hotel, among packed tables of tourists and businesspeople, with a buffet of delicacies on long tables just inside open doors, I watched her face switch among the radio stations of memory, then she quietly sang a song she'd heard over sixty years before, one a handsome young soldier had sung to her as she walked past:

Ty jeszcze o tym nie wiesz dziewczyno,
Ze od niedawna jestés przyczyną,
Mych snów, pięknych snów,
Ja mógłbym tylko wziąść cię na ręce,
I jeszcze więcej niż dzziś,
Kochać cię.

You don't know about this yet, my girl,
That lately you have been the cause
Of my dreams, beautiful dreams.

If I could only lift you up in my arms,
And, even more than today,
Love you.

Halina's face flushed a little from that tall umbrella drink of memory, stored among more tragic images, as wartime memories often are, having their own special filing system, their own ecology. If other diners overheard, no one gave a sign, and as I looked around the archipelago of tables, I realized that out of fifty or so people, she was the only one old enough to harbor wartime memories.

Ryś, a civil engineer and a father himself, lives in downtown Warsaw today in an eight-story walk-up, minus pets. "A dog couldn't climb the stairs!" he explained as we lurched from landing to landing. Tall and slender, in his seventies, he appears fit from all the stair-climbing, friendly and hospitable, but also a little wary, not surprising given the war lessons ingrained from an early age. "We lived from moment to moment," Ryś said, sitting in his living room, watched over by photographs of his parents, many of their books, a framed drawing of a forest bison, and a sketch of his father. Zoo life hadn't seemed at all unusual to him as a boy, he said, because "it was all I knew." He told of watching a bomb fall near the villa and realizing that he was close enough to be killed, had it gone off. He remembered posing for Magdalena Gross, sitting for long hours while she coaxed clay, existed in it really, and he relished her chirpy attentions. I learned from him that his mother filled the villa's upstairs terrace with overflowing flower boxes in warm weather, that she especially liked pansies, the flowers with pensive faces (from the French *pensée*), that she preferred the music of Chopin, Mozart, and Rossini. No doubt he found some of my questions odd—I hoped to learn about his mother's scent, how she walked, her gestures, her tone of voice, how she wore her hair. To all such inquiries, he answered "average" or "normal," and I soon realized those were memory traces he either didn't visit or didn't wish to share. His sister Teresa, born late in the war, married and lives in Denmark. I invited grown-up Ryszard to visit the villa with me, and he kindly obliged. As we explored his childhood home, stepping carefully over the doorframes with decorative anvil-shaped thresholds, I was struck by the way he tested his memory, often comparing what is to what was in much the same way Antonina described him doing as a boy, when they returned to the bombed zoo at the end of the war.

In one of those twists of fate that pepper history, the Berlin Zoo was heavily bombed, just as the Warsaw Zoo had been, assailing Lutz Heck with many of the same concerns and hardships he'd imposed on the Żabińskis. In his autobiography,

Animals—My Adventure, he writes movingly about his fatally wounded zoo. Unlike the Żabińskis, he knew exactly what devastation to expect, having witnessed it firsthand in Warsaw, whose zoo bombing he never mentions. His safari animals, large collection of photographs, and numerous diaries vanished by war's end. As the Soviet army advanced, Lutz left Berlin to avoid being arrested for looting Ukrainian zoos, and he spent the rest of his life in Wiesbaden, making hunting trips abroad. Lutz died in 1982, a year after his brother Heinz. Lutz's son Heinz immigrated to the Catskills in 1959, where he ran a small zoo famous for its herd of Przywalski horses, descended from those nurtured by Heinz Heck throughout the war. At one point, the Munich Zoo had assembled the largest herd of Przywalski horses outside of Mongolia (some stolen from the Warsaw Zoo).

In all, around three hundred people passed through the way station of the Warsaw Zoo, en route to the rest of their nomadic lives. Jan always felt, and said publicly, that the real heroine of this saga was his wife, Antonina. "She was afraid of the possible consequences," he said to Noah Kliger, who interviewed him for the Israeli newspaper *Yediot Aharonot*, "she was terrified the Nazis would seek revenge against us and our young son, terrified of death, and yet she kept it to herself, and helped me [with my Underground activities] and never ever asked me to stop."

"Antonina was a housewife," he told Danka Narnish, of another Israeli paper, "she wasn't involved in politics or war, and was timid, and yet despite that she played a major role in saving others and never once complained about the danger."

"Her confidence could disarm even the most hostile," he

told an anonymous reporter, adding that her strength stemmed from her love of animals. "It wasn't just that she identified with them," he explained, "but from time to time she seemed to shed her own human traits and *become* a panther or a hyena. Then, able to adopt their fighting instinct, she arose as a fearless defender of her kind."

To reporter Yaron Becker, he explained: "She had a very traditional Catholic upbringing and that didn't deter her. On the contrary it strengthened her determination to be true to herself, to follow her heart, even though it meant enduring a lot of self-sacrifice."

Intrigued by the personality of rescuers, Malka Drucker and Gay Block interviewed over a hundred, and found they shared certain key personality traits. Rescuers tended to be decisive, fast-thinking, risk-taking, independent, adventurous, openhearted, rebellious, and unusually flexible—able to switch plans, abandon habits, or change ingrained routines at a moment's notice. They tended to be nonconformists, and though many rescuers held solemn principles worth dying for, they didn't regard themselves as heroic. Typically, one would say, as Jan did: "I only did my duty—if you can save somebody's life, it's your duty to try." Or: "We did it because it was the right thing to do."

Antonina died in 1971, her husband three years later.

Chapter 36

I. Białowieża, 2005

At the edge of a primeval forest in northeastern Poland, time seems to evaporate, as two dozen horses graze on the marsh grass beneath colossal pine trees and a dazzling blue sky. On frosty mornings they browse inside bubbles of steam and leave a sweet leathery odor behind them when they go. Their body-fog travels with them, but their scent can remain for hours as invisible clouds above jumbled hoofprints, and some-times, on a gravel path or leaf-strewn trail, where no hoof-prints tell, one enters a pocket of gamy air and is suddenly surrounded by essence of wild horse.

Spring through fall, the horses live unaided by humans, wading in the ponds and grazing on bushes, tree branches, algae, and grass. Snow falls in mid-October and remains until May. In winter, they hungrily paw through the snow to find dry grass or rotting apples, and rangers of the mounted Horse Guard sometimes provide hay and salt. Blessed with bolt-and-leap muscle, they have little fat to insulate them on icy days, so they grow shaggy coats that mat easily. It's then that they most resemble the horses painted on the cave walls at prehistoric sites throughout the Loire Valley.

How startling it is to shed the here and now and watch

what might be ancient horses browsing at the forest-edge meadows just as humans did millennia ago. They're strikingly beautiful creatures: dun with a black stripe down the back and a dark mane (sometimes a foal will be born with black face and fetlocks and a zebra-striped leg or two). Although they have long ears and large thick necks, they're lightly built and fast. Unlike domestic horses, they turn white in winter, just as ermine and arctic hares do, blending in with the landscape. Then ice clots like marbles in coarse manes and tails, and as they stomp their hooves build up platforms of snow. Still, they thrive on harsh weather and poor diet; and although the stallions battle fiercely, with bared teeth and thwacking necks, they heal fast as a shaman's spell. "In a world older and more complete than ours they move," Henry Beston writes of wild animals in *The Outermost House*, "gifted with extensions of the senses we have lost or never attained, living by voices we shall never hear."

In Białowieża, one can also find re-created aurochsen, a favorite game animal of Julius Caesar, who described them to friends back in Rome as savage black bulls "a little less than elephants in size," strong and fast. "They spare neither man nor beast," he wrote. "They cannot be brought to endure the sight of men, nor be tamed, even when taken young." Apparently, men of the Black Forest trained rigorously to hunt aurochsen bulls (the cows were left to breed), and those "who have killed a great number—the horns being publicly exhibited in evidence of the fact—obtain great honour. The horns . . . are much sought after; and after having been edged with silver . . . are used for drinking vessels at great feasts." Some of those silver-tipped horns remain in museums. But by 1627, the last true aurochsen had been killed.

Yet here tarpans, bison, and aurochsen apparently graze, roaming the closely guarded nature preserve on the Polish-Belorussian border that's been a royal favorite since the 1400s, a realm of magic and monsters that inspired many European fairy tales and myths. King Kazimierz IV found it so enchanting that he spent seven years (1485–92) living in a simple forester's lodge and managing affairs of state from his sylvan home.

What's so awe-inspiring about this landscape that it could bewitch people from many cultures and eras, including Lutz Heck, Göring, and Hitler? For starters, it contains five-hundred-year-old oak trees, as well as soaring pine, spruce, and elm rising like citadels hundreds of feet tall. It boasts 12,000 species of animal, from one-celled protozoa to such large mammals as boar, lynx, wolf, and moose; and, of course, there are its bands of throwback aurochsen, tarpans, and bison. Beavers, martens, weasels, badgers, and ermine glide through the marshes and ponds, while Pomeranian eagles share the skies with bats, goshawks, tawny owls, and black storks. On any given day, one is likely to encounter more elk than humans. The air smells of balsam and pine needles, sphagnum moss and heather, berries and mushrooms, marshy meadows and peat bogs. Small wonder Poland has chosen the preserve as its only natural national monument, and that it also merits the honor of being a World Heritage Site.

Because the preserve is closed to hunters, loggers, and motorized vehicles of any sort, it's the last refuge of unique flora and fauna, and for that reason, park rangers guide tiny groups of hikers along designated paths, where they're forbidden to litter, smoke, or speak above a whisper. Nothing may be removed, not even a leaf or stone as souvenir. All traces of humankind, especially noise, are discouraged, and rangers use rubber-tired horsecarts in the park; they saw up fallen trees by hand and use horses to drag them out.

In what's known as the "strict preserve," one sees many fallen, dead, and decaying trees which, oddly enough, create the backbone and great strength of the forest, which is why activists vigorously defend its dead wood. Wind-thrown trees, fallen naturally and rotting, provide home to a throng of creatures: 3,000 species of fungi, 250 species of mosses, 350 species of lichen, 8,791 species of insects, mammals, and birds. Guides and a museum with dioramas teach the park's ecology and history, but few visitors realize how richly it appealed both to Nazi racism and romanticism.

As twilight descends on the marshes of Białowieża, hundreds of starlings fly up all at once and create a great funnel, then the flock descends to find shelter for the night among the pond grasses. I'm reminded of Antonina's love of starlings, and of Magdalena "the Starling," and of Lutz Heck, too, who fancied "the little iridescent green glossy starling [which] warbled its little song with wide open beak, its small body literally shaking under the force of its notes." The eugenics and breeding experiments that thrived with Heck's ambitions, Göring's lust for game, and Nazi philosophy, in the end, ironically, helped to save scores of rare plants and endangered animals.

Understandably bitter about the Hecks' Nazi ties and motives, some Polish patriots were (and still are) quick to contend that these animals may resemble their ancient ancestors but are, nonetheless, technically counterfeits. Cloning wasn't available in the days of the Heck brothers, or they would surely have mastered it. Some zoologists, who prefer to call them "near tarpans" and "near aurochsen," associate them with political agendas. The horses, "although not truly wild animals, are big, exotic creatures which have a history coloured with drama, dedication and skullduggery," biologist Piotr Daszkiewicz and journalist Jean Aikhenbaum declare in *Aurochs, le retour . . . d'une supercherie nazie* (1999). They paint the Hecks as con men who staged a colossal Nazi hoax—by creating a new species, not resurrecting an extinct one. Herman Reichenbach, reviewing their book in *International Zoo News*, counters that Daszkiewicz and Aikhenbaum's book is short on facts and essentially "what the French call a *polémique* . . . and the Americans a 'hatchet job' . . . [but] [p]erhaps the Hecks deserve it; after the war both were less than candid about their association with the Nazi dictatorship. . . . [R]ecreating an ancient Germanic environment (within park borders) was as much Nazi ideology as getting back Alsace again."

However, Reichenbach envisions an important role for the Hecks' creations: "They can still help preserve a natural environment of mixed forest and meadows. . . . And as a feral type of cattle, the aurochsen may also be able to enhance the gene pool of a domestic animal that has become impoverished genetically during the last decades. Attempting to back-breed the aurochs may have been a folly, but it was not a crime." Professor Z. Pucek of the Białowieża Nature Preserve denounces the Heck cattle as "the biggest scientific swindle of the twentieth century." And so the controversy continues, debated in journals and online, with a passage from American C. William Beebe often cited. In *The Bird: Its Form and Function* (1906), Beebe writes: "The beauty and genius of a work of art may be reconceived, though its first material expression be destroyed; a vanished harmony may yet again inspire the composer; but when the last individual of a race of living things breathes no more, another heaven and another earth must pass before such a one can be again."

There are many forms of obsession, some diabolical, some fortuitous. Strolling through Białowieża's mass of life, one would never guess the role it played in Lutz Heck's ambitions, the Warsaw Zoo's fate, and the altruistic opportunism of Jan and Antonina, who capitalized on the Nazis' obsession with prehistoric animals and a forest primeval to rescue scores of endangered neighbors and friends.

II.

Warsaw today is a spacious green city with acres of sky, in which tree-lined avenues flow down to the river, ruins mix with new trends, and everywhere tall old trees offer scent and shade. In the zoo district, Praski Park still teems with cloyingly sweet lindens and, in summer, honeying bees; and across the river, where the Jewish Ghetto once stood, a park of chestnut trees surrounds a plaza and stark monument. After the defeat of Communism in 1989, with characteristic humor the Poles turned the former Gestapo headquarters into the Ministry of Education, the former KGB headquarters into the Ministry of Justice, the Communist Party headquarters into the Stock

Exchange, and so on. But the architecture of Old Town is a visual hymn, rebuilt after the war in Vistula Gothic, based on old drawings and paintings by seventeenth-century Venetian Bernardo Bellotto—a feat organized by Emilia Hizowa (who invented Zegota's push-button sliding walls). Some buildings show recycled rubble from the bombed city embedded in their façades. Dozens of statues and monuments grace Warsaw's streets, because Poland is a country half submerged in its heavily invaded past, fed by progress, but always partly in mourning.

Retracing Antonina's footsteps from the downtown apartment where she lodged with relatives during the siege of Warsaw, I walked to Miodowa Street, crossed the old moat, and slipped through the crumbling brick walls encircling Old Town. As one enters a world of tightly knit row houses, shoes slide over cobblestones and the body continuously balances itself in tiny increments until the stones grow larger, smoothed by centuries of footsteps. In rebuilding the city after the war, planners used as many of the original stones as possible, and in *The Street of Crocodiles*, Antonina's contemporary Bruno Schulz describes the same colorful mosaic underfoot that exists today: "some of the pale pink of human skin, some golden, some blue-gray, all flat, warm and velvety in the sun, like sundials trodden to the point of obliteration, into blessed nothingness."

On such narrow streets, electric streetlamps (once gas) sprout from corner buildings and double-sashed windows stand open like an Advent calendar. Black stovepipe gutters underscore the terra-cotta rooftops, and some of the painted stucco walls have chipped to reveal the flesh-red brick foundations underneath.

I turned down Ulica Piekarska (Bakers Street), as cobblestones fanned and eddied like a petrified creek bed, then turned left into Piwna (Beer) Street, past a shrine recessed into the second story of a housefront and filled with a wooden saint flanked by floral offerings. Next I passed Karola Beyera, the coin collectors' club, and three short wooden doors leading into courtyards, then turned left at the pyramidal wall of a corner building, and finally entered the large open Market Square. In the early days of the war, when Antonina shopped there, few vendors risked setting up, the amber and antiques stores stayed closed, patrician homes locked their doors, and the fortune-telling parrot of the 1930s was nowhere to be seen.

Leaving the square, I strolled toward the old fortifications to visit the closest well, following a wall of sooty bricks that curves around to medieval towers with funnel-capped lookouts and narrow slits that once hid archers. In summer, the mock orange trees along this walk froth with white flowers visited by fat black-and-white magpies. Floating above the wall, the canopies of crab apple trees scramble for sun. On Rycerka (Knight) Street, I reached a small square and a black pillar emblazoned with a mermaid wielding a sword—Warsaw's symbol. It's a chimera I think Antonina would have identified with: a defender half woman, half animal. On both sides of the pillar, a bearded god spills water from his mouth, and it's easy to picture Antonina setting down her basket, angling a jug under a spout, and waiting as life gurgled up from the earth.

Details

Chapter 2

Thievery posed another worry (17): A few years back, burglars broke into the Warsaw Zoo's aviary and stole various owls, a raven, and a condor, and officials assumed they nabbed the owls and raven to mislead, their real target being the condor, whose black-market value had soared. On another occasion a robber stole a baby penguin. Zoo abductions happen everywhere, usually commissioned by breeders or laboratories, but sometimes by individual collectors. Notably, a beautiful cockatoo stolen from the Duisburg Zoo was later found dead and stuffed, in the apartment of a couple who had received it as an anniversary present.

a Burmese man invented . . . a hopping stick (18): Pogo sticks, all the rage of the 1920s, were actually patented by the American George Hansburg.

backward-bent red knees (18): Flamingos look like they have backward-facing knees, but those are actually their ankles. Their knees float higher up, hidden by feathers.

Chapter 3

The Żabińskis' country cottage [in Rejentówka] (26): Many of these details come from Helena Boguszewska, who owned a neighboring property.

Chapter 4

scenes of the dead and dying etched into memory (36): Antonina's recollection is matched by that of Wiktor Okulicz-Kozaryn, a retired engineer, who watched the same scene as a boy, and remembers "German aircraft flying low over the crowd, shooting and killing many people . . . [and] two Polish planes attacking a German bomber above a field, the plane flaming, then one parachute floating down near some trees."

"That which doesn't kill you, makes you stronger" (36): Friedrich Nietzsche, *The Twilight of the Idols* (1899).

Chapter 5

the newly invented jukebox (42): Jukeboxes were invented in the 1930s, to supply music in back-road *jooks*—Carolina creole for joints that were a combination of bawdy house, gambling den, and dance shack.

Chapter 6

"While speaking to you now, I can see it through the window in its greatness and glory" (51): Stefan Starzyński, quoted in *Warsaw and Ghetto* (Warsaw: B. M. Potyralsey, 1964).

"I rely on the population of Warsaw . . . to accept the entry of the German forces" (54): Rommel quoted in Israel Gutman, *Resistance: The Warsaw Ghetto Uprising* (New York: Houghton Mifflin, 1994), p. 12.

"ruthlessly exploit this region as a war zone and booty country" (57): *Proceedings of the Trial of the Major War Criminals Before the International Military Tribunal, Nuremberg,* vol. 290, ND 2233-PS; quoted in Anthony Read, *The Devil's Disciples: Hitler's Inner Circle* (New York: W. W. Norton, 2004), p. 3.

over a span of five years (57): Adam Zamoyski, *The Polish War: A Thousand Year History of the Poles and Their Culture* (New York: Hippocrene Books, 1994), p. 358.

"from the very beginning, I was connected to the Home Army" (58): Jan Żabiński quoted in a Yiddish newspaper, in Israel, on the occasion of the Żabińskis being honored by Yad Vashem as "Righteous Among Nations." Newspaper article provided by Ryszard Żabiński.

Chapter 7

director of the Munich Zoo (62): Heinz Heck became director of the Hellabrunn Zoo in Munich in 1928, where he remained until 1969.

Esperanto (63): Esperanto was invented in 1887 in Białystok by Dr. Ludovic Lazar Zamenhof, an eye doctor, who chose the pseudonym of "Doktoro Esperanto" (Dr. Hope). Immersed in Białystok's polyglot world, he noted how much distrust and misunderstanding between ethnic groups stemmed from language barriers, so he designed a neutral lingua franca.

Chapter 8

"The campfire flickering in front of me" (66): Lutz Heck, *Animals—My Adventure,* trans. E. W. Dickies (London: Methuen, 1954), p. 60.

back-breeding project (69): Though Polish scientist Tadeusz Vetulani had tried the same back-breeding process years before without success, Heck stole Vetulani's research and, ultimately, thirty animals, which he sent to Germany, later installing them in Rominten and then Białowieża.

biological aims of the Nazi movement (70): Much as Hitler publicly championed a fit, vigorous Aryan race, Goebbels had a clubfoot, Göring was obese and addicted to morphine, and Hitler himself seems to have been suffering from third-stage syphilis by the end of the war, addiction to uppers and downers, and quite possibly Parkinson's. Hitler's doctor, Theo Morell, a renowned specialist in syphilis, accompanied him everywhere, syringe and gold-foil-wrapped vitamins at the ready. Rare film footage shows Hitler using his steady right hand to shake hands with a line of boys, while his left, hidden behind his back, displays Parkinson's distinctive tremor.

What were his so-called vitamins? According to criminologist Wolf Kemper (*Nazis on Speed: Drogen im 3. Reich* [2002]), the Wehrmacht commissioned an array of drugs that would increase focus, stamina, and risk-taking, while reducing pain, hunger, and fatigue. Between April and July of 1940, troops received over 35 million 3-milligram doses of the addictive and mood-altering amphetamines Pervitin and Isophan.

In a letter dated May 20, 1940, twenty-two-year-old Heinrich Böll, then stationed in occupied Poland, despite his "unconquerable (and still unconquered) aversion to the Nazis," wrote his mother in Cologne to rush him extra doses of Pervitin, which German civilians were buying over the counter for their own use. (Leonard L. Heston and Renate Heston, *The Medical Casebook of Adolf Hitler* [London: William Kimber, 1979], pp. 127–29.)

Josef Mengele (71): Josef Mengele grew up in a family of Bavarian industrialists, and declared his religion as Catholic on official forms (instead of "believer in God," as Nazism preferred). Genetic abnormalities fascinated him, and "Dr. Auschwitz," as he came to be known, had an ample pool of children on whom to do experiments which the Frankfurt court would later denounce as "hideous crimes" performed "willfully and with bloodlust," which often included vivisection or murder. "He was brutal but in a gentlemanly, depraved way," one prisoner reported, and others described him as "very playful," "a Rudolph Valentino type," always smelling of eau de cologne. (Quoted in Robert Jay Lifton, *The Nazi Doctors: Medical Killing and the Psychology of Genocide* [New York: Basic Books, 1986], p. 343.) "In selecting for death or in killing people himself, the essence of Mengele was flamboyant detachment—one might say disinterestedness—and efficiency," Lifton concludes (p. 347).

As new prisoners arrived, guards marched up and down the lines calling out *"Zwillinge, zwillinge!"* as they hunted twins for Mengele to tamper with in gruesome ways. Changing eye color became his favorite line of research, and on one wall of his office, he displayed an array of surgically removed eyes pinned up like a collection of moths.

"a deliberate, scientifically founded race policy" (72): Konrad Lorenz quoted in Ute Deichmann, *Biologists Under Hitler,* trans. Thomas Dunlap (Cambridge, Mass.: Harvard University Press, 1996), p. 187.

"the healthy volkish body often does not 'notice' how it is being pervaded by elements of decay" (72): Konrad Lorenz, "Durch Domestikation verursachte Störungen artewigen Verhaltens," *Zeitschrift für angewandte Psychologie und Charakterkunde,* vol. 59 (1940), p. 69.

Hermann Göring (72): As part of Hitler's inner circle, he quickly rose to "air minister," as well as "Master of the German Hunt" and "Master of German Forests." More than just an avid huntsman—he once had a stag from his estate flown to him in France so that he could track and shoot it—Göring identified hunting with life at his boyhood castle, and dreamt of returning Germany to its lost greatness ("Our time will come again!" he would proclaim). Weekends he spent in the forests, and seizing any excuse to combine politics and hunting, he hosted haute cuisine shooting parties. Hitler didn't hunt, though he often wore hunter's garb, especially at his lodge in the Alps, as if at any moment he might release a falcon or leap into the saddle and chase a candelabra-horned stag.

Fascinated by boar hunting, Göring prized a custom-made fifty-inch boar spear with a leaf-shaped blade of blue steel, a dark mahogany grip, and a steel shaft with two hollow pleated spheres that rattled to scare his prey from the underbrush. Göring took dozens of hunting trips with friends, foreign dignitaries, and members of the German high command from the mid-1930s to late 1943; and documents show that even in January and February of 1943, while Germany was losing on the Russian front, Göring was at his castle, hunting Rominten wild boar and Prussian royal stags. (During this same period he also introduced ballroom dancing lessons for Luftwaffe officers.)

Chapter 9

So many excellent books have been written about daily life in the Ghetto, the Jew roundups, and the horrors of the death camps, that I don't linger on them. A particularly vivid account of the Uprising that comes to mind is *A Fragment of the Diary of the Rubbish Men,* by Leon Najberg, who fought with armed stragglers among the ruins until the end of September.

European Bison Stud Book (76): It continues to this day, though it's now issued in Poland. No bloodline information is kept on the wild bison, which rangers simply keep an eye on and count.

For good discussions of the motif, see Piotr Daszkiewicz and Jean Aikhenbaum, *Aurochs, le retour . . . d'une supercherie nazie* (Paris: HSTES, 1999), and Frank Fox, "Zagroz˙one gatunki: ˙Zydzi i z˙ubry" (Endangered Species: Jews and Buffalo), *Zwoje*, January 29, 2002.

"I've been asked that a lot" (77): Heck, *Animals,* p. 89.

Chapter 10

any illness that kills one animal threatens to wipe out all (78): This curse of closely knit species also applies to our dairy cows, now almost clones of one another; an illness that kills one can kill all.

A 2006 study of mitochondrial DNA (78): "The Matrilineal Ancestry of Ashkenazi Jewry: Portrait of a Recent Founder Event": Doron M. Behar, Ene Metspalu, Toomas Kivisild, Alessandro Achilli, Yarin Hadid, Shay Tzur, Luisa Pereira, Antonio Amorim, Lluís Quintana-Murci, Kari Majamaa, Corinna Herrnstadt, Neil Howell, Oleg Balanovsky, Ildus Kutuev, Andrey Pshenichnov, David Gurwitz, Batsheva Bonne-Tamir, Antonio Torroni, Richard Villems, and Karl Skorecki. *American Journal of Human Genetics,* March 2006.

some say to a man, some a woman (79): That person didn't live all alone on the planet; it's simply that no one else's offspring survived.

"Germany's crime is the greatest crime the world has ever known" (79): Pierre Lecomte du Noüy, *La dignité humaine* (1944).

the Underground, whose foothold in the Praga district in time reached 90 platoons with 6,000 soldiers (80): Norman Davies, *Rising '44: The Battle for Warsaw* (London: Pan Books, 2003), p. 183.

Chapter 11

Hitler had ordered (86): From a transcript read at the Nuremberg trials, reported in "The Fallen Eagles," *Time*, December 3, 1945.

all Poles drew punishment (87): Out of its prewar population of 36 million, Poland lost 22 percent, more than any other country in Europe. After the war, Yad Vashem in Jerusalem, the State Tribunal of Israel, detailed some of Christian Poland's ordeal, and how, in addition to the 6 million Jews killed, 3 million Catholics died, "but what is even worse, it lost especially its educated classes, youth and any elements which could in the future oppose one or the other of the two totalitarian regimes. . . . According to the German plan, Poles were to become a people without education, slaves for the German overlords."

Chapter 12

"constant tense clamor" (93): Michał Grynberg, ed., *Words to Outlive Us: Eyewitness Accounts from the Warsaw Ghetto*, trans. Philip Boehm (London: Granta Books, 2003), p. 46.

At one point Himmler invited Werner Heisenberg to establish an institute to study icy stars because, according to the cosmology of *Welteislehre*, based on the observations of the Austrian Hanns Hörbiger (author of *Glazial-Kosmogonie* [1913]), most bodies in the solar system, our moon included, are giant icebergs. A refrigeration engineer, Hörbiger was persuaded by how shiny the moon and planets appeared at night, and also by Norse mythology, in which the solar system emerged from a gigantic collision between fire and ice, with ice winning. Hörbiger died in 1931, but his theory became popular among Nazi scientists and Hitler swore that the unusually cold winters in the 1940s proved the reality of *Welteislehre*. Nicholas Goodrick-Clarke's *The Occult Roots of Nazism* explores the influence of such magnetic lunatics as Karl Maria Wiligut, "the Private Magus of Heinrich Himmler," whose doctrines influenced SS ideology, logos, ceremonies, and the image of its members as latter-day Knights Templars and future breeding stock for the coming Aryan utopia. To this end, Himmler founded Ahnenerbe, an institute for the study of German prehistory, archaeology, and race, whose staff wore SS uniforms. Himmler also acquired Wewelsburg Castle in Westphalia to use immediately for SS education and pseudoreligious ceremonies, and remodel into a future site altogether more ambitious, "creating an SS vatican on an enormous scale at the center of the millenarian greater Germanic Reich."

"In Warsaw the Ghetto was no longer anything but an organized form of death" (93): Michael Mazor, *The Vanished City: Everyday Life in the Warsaw Ghetto,* trans. David Jacobson (New York: Marsilio Publishers, 1993), p. 19.

"mere mention of a threat" (93): Grynberg, *Words to Outlive Us,* pp. 46–47.

Chapter 14

"just to facilitate your research, [and] to let you know what we think of you" (109): *After Karski,* p. 267, quoted in Davies, *Rising '44,* p. 185.

Chapter 15

president of Warsaw (113): The president of Warsaw is equivalent to the mayor of a city.

any excuse to visit friends "to keep up their spirits and smuggle in food and news" (115): See Jan E. Rostal, "In the Cage of the Pheasant," *Nowiny i Courier,* October 1, 1965.

"a period of political repression, censorship, and infringement on personal liberties" (117): Milton Cross, *Encyclopedia of the Great Composers and Their Music,* Doubleday, 1962, pp. 560–61.

the Warsaw Ghetto's Labor Bureau (118): Workers deported to Germany by the Arbeitsamt had to wear a purple *P* on their sleeve, and were denied church, cultural pursuits, and public transportation. Sex with a German was punishable by death. (Davies, *Rising '44,* p. 106)

"when he saw the beautiful beetles and butterflies, he forgot all about the world" (122): *Polacy z pomocą Żydom* (Poles

Helping Poles), 2nd edition (Kraków: Wydawnictwo Znak, 1969), pp. 39–45.

"The creation, existence, and destruction of the Ghetto" (124): Philip Boehm, introduction to Grynberg, *Words to Outlive Us*, p. 3.

"Frankenstein was a short, bull-legged, creepy-looking man" (125): Jack Klajman with Ed Klajman, *Out of the Ghetto* (London: Vallentine Mitchell, 2000), pp. 21, 22.

Chapter 16

"I wanted to tell Jan—'Let's run.'" (137): Lonia Tenenbaum, in *Polacy z pomocą Żydom* (Poles Helping Poles).

about half of the full collection, which Jan told a journalist ran to four hundred boxes (138): Rostal, "In the Cage of the Pheasant."

Chapter 17

"doctrine of blood and soil" (140): Karl Friederichs quoted in Deichmann, *Biologists Under Hitler*, p. 160.

***Epidemics Resulting from Wars* (141):** Friedrich Prinzing, *Epidemics Resulting from Wars* (Oxford: Clarendon Press, 1916).

"Antisemitism is exactly the same as delousing" (141): speech to SS officers, April 24, 1943, Kharkov, Ukraine; reprinted in United States Office of Chief of Counsel for the Prosecution of Axis Criminality, *Nazi Conspiracy and*

Aggression (Washington, D.C.: United States Government Printing Office, 1946), vol. 4, pp. 572–78, 574.

the slogan "JEWS—LICE—TYPHUS" (142): report by Ludwig Fischer quoted in Gutman, *Resistance,* p. 89.

he was "seized by the wish not to have a face" (142): Hannah Krall, *Shielding the Flame: An Intimate Conversation with Dr. Marek Edelman, the Last Surviving Leader of the Warsaw Ghetto Uprising* (New York: Henry Holt, 1977), p. 15.

"when the three horsemen of the Apocalypse" (143): Stefan Ernest quoted in Grynberg, *Words to Outlive Us,* p. 45.

"When you eat and drink" (144): Alexander Susskind, quoted in Daniel C. Matt, ed., *The Essential Kabbalah: The Heart of Jewish Mysticism* (San Francisco: HarperCollins, 1995; translated from Dov Baer, Maggid Devarav l'Ya'aquov), p. 71.

"One hears the [Teaching's] voice" (146): Nehemia Polen, *The Holy Fire: The Teachings of Rabbi Kalonymus Kalman Shapira, the Rebbe of the Warsaw Ghetto* (Lanham, Md.: Rowman & Littlefield, 1994), p. 163.

one Ghetto inhabitant (146): Marek Edelman in Krall, *Shielding the Flame.* After the war Edelman became a cardiologist, commenting that "when one knows death so well, one has more responsibility for life."

Chapter 18

"The personality of animals will develop" (152): postwar interview by Danka Harnish, in Israel, translated from Hebrew by Haviva Lapkin of the Lorraine and Jack N. Friedman Commission for Jewish Education, West Palm Beach, Florida, April 2006.

"Consisting of 28,000 Jews" (160): Gunnar S. Paulsson, *Secret City: The Hidden Jews of Warsaw, 1940–1945* (New Haven, Conn.: Yale University Press, 2002), p. 5.

"Tenants visited each other" (161): Alicja Kaczyńska, *Obok piekła* (Gdańsk: Marpress, 1993), p. 48; quoted in Paulsson, *Secret City,* pp. 109–10.

Chapter 20

"Uncle is planning (God preserve us) to hold a wedding" (169): from Ruta Sakowska, ed., *Listy o Zagladzie* (*Letters About Extermination*) (Warsaw: PWN, 1997). Jenny Robertson, *Don't Go to Uncle's Wedding: Voices from the Warsaw Ghetto* (London: Azure, 2000).

"district of the damned" (171): Janusz Korczak, *Ghetto Diary* (New Haven, Conn.: Yale University Press, 2003), p. x.

"adhesions, aches, ruptures, scars" (172): *Ghetto Diary,* p. 9.

"Thank you, Merciful Lord" (172): *Ghetto Diary,* p. 8.

"When I collect the dishes myself" (173): *Ghetto Diary,* p. 107.

whose pure souls make possible the world's salvation (174): Betty Jean Lifton, introduction to *Ghetto Diary,* p. vii.

Chapter 21

"The people of Zegota were not just idealists but activists, and activists are, by nature, people who know people" (176): Irene Tomaszewski and Tecia Werbowski, *Zegota: The Rescue of Jews in Wartime Poland* (Montreal, Canada: Price-Patterson Ltd., 1994).

70,000–90,000 people (177): from Paulsson, *Secret City,* p. 163.

a giant jar Jan had once used in a cockroach study (183): Jan Żabiński, "The Growth of Blackbeetles and of Cockroaches on Artificial and on Incomplete Diets," *Journal of Experimental Biology* (Company of Biologists, Cambridge, UK), vol. 6 (1929): pp. 360–86.

Chapter 23

Surprisingly, sketchy telephone service continued (190): Emanuel Ringelblum, *Polish-Jewish Relations During the Second World War* (New York: Howard Festig, 1976), pp. 89–91.

"a husk or shell that has grown up around a spark of holiness, masking its light" (191): Michael Wex, *Born to Kvetch: Yiddish Language and Culture in All of Its Moods* (New York: St. Martin's Press, 2005), p. 93.

Yiddish's famous curses . . . "May you piss green worms!" (197): Wex, *Born to Kvetch,* pp. 117, 132, 137.

"Here a terrible depression reigns" (198): Judit Ringelblum, *Beit Lohamei ha-Getaot* (Haifa, Israel: Berman Archives); quoted in Paulsson, *Secret City,* p. 121.

Chapter 24

"For him, I would do anything," he once told a friend. "Believe me, if Hitler were to say I should shoot my mother, I would do it and be proud of his confidence" (200): Otto Strasser, *Mein Kampf* (Frankfurt am Main: Heinrich Heine Verlag, 1969), p. 35.

"Nearby, on the other side of the wall, life flowed on as usual" (201): Cywia Lubetkin, *Extermination and Uprising* (Warsaw: Jewish Historical Institute, 1999); quoted in Robertson, *Don't Go to Uncle's Wedding,* p. 93.

"the Germans have removed, murdered, or burned alive tens of thousands of Jews" (202): Stefan Korboński, *Fighting Warsaw: The Story of the Polish Underground State, 1939–1945* (New York: Hippocrene Books, 2004), p. 261.

Chapter 25

"There I saw a dozen more or less undressed ladies" (210): From the account of Władysław Smólski in *Righteous Among Nations: How Poles Helped the Jews, 1939–1945,* edited by Władysław Barloszewski and Zofia Lewin (London: Earlscourt Publications Ltd., 1969), pp. 255–59.

operated on Jewish men to restore foreskins (212):
Schultheiss, Dirk, M.D., et al., "Uncircumcision: A Historical
Review of Preputial Restoration," *Plastic and Reconstructive
Surgery,* vol. 101, no. 7 (June 1998): pp. 1990–98.

"Suffering took hold of me" (213): a personal reminiscence
deposited with the Jewish Historical Institute after World
War II, and published in *Righteous Among Nations,* p. 258.

Chapter 27

"man as a sensitive receiver" (228): Goodrick-Clark, *The
Occult Roots of Nazism,* p. 161.

Chapter 28

**"It is the imaginary perils, [the] supposed observation by
the neighbour, porter, manager, or passer-by" (233):**
Ringelblum, *Polish-Jewish Relations,* p. 101.

"A picture falling off a wall" (234): Sophie Hodorowicz Knab,
Polish Customs, Traditions, and Folklore (New York: Hippocrene
Books, 1996), p. 259.

people "walking on quicksand" (236): Janina in *Righteous
Among Nations,* p. 502.

"I am lucky . . . I can do wonders" (236): Rachela "Aniela"
Auerbach, postwar testimony in *Righteous Among Nations,* p.
491.

"the perennial protector of the underdog" (237): Basia in *Righteous Among Nations,* p. 498.

In a postwar interview with London's *White Eagle-Mermaid* (240): May 2, 1963.

Chapter 29

"If I maintain my silence about my secret" (247): Arthur Schopenhauer, *Parerga and Paralipomena,* trans. E. F. J. Payne (New York: Oxford University Press, 2000), vol. 1, p. 466 (chap. 5, "Counsels and Maxims").

Chapter 30

blanket-bomb German cities, including Dresden (258): In the ensuing firestorm, counting victims became impossible, though it's now estimated that 35,000 people perished in Dresden. The rare manuscripts of eighteenth-century Italian composer Tomaso Albinoni, whose Adagio in G Minor has become synonymous with mournfulness, also vanished in flames.

the most superstitious of cultures (259): Many Poles believed in signs and witchery. It was once common for Warsawians to read their fate in a deck of regular (not tarot) cards, or predict the future, especially marriage, by melting wax on a spoon and slowly pouring it into a bowl of cold water. Supposedly, the shape the wax took revealed one's fate—a hammer or helmet shape told a boy he'd be soldiering soon, and a girl that she'd marry a blacksmith or soldier. If a girl dripped wax resembling a cabinet or other furniture, she'd marry a carpenter; if it looked more like wheat or a wagon, she'd marry a farmer. A violin or trumpet meant the person would become a musician.

According to Polish lore, Death appears to humans as an old woman in a white winding sheet carrying a scythe, and dogs can easily spot her. So one can glimpse Death "by stepping on a dog's tail and looking between his ears."

Chapter 31

Russians (263): The wild-eyed Russian soldiers, known as "Wlasowcy," were soldiers of the Russian general Wlasow, who was collaborating with the Third Reich.

"The tram-cars were crowded with young boys" (264): Stefan Korboński, *Fighting Warsaw: The Story of the Polish Underground State, 1939–1945,* trans. F. B. Czarnomski (New York: Hippocrene Books, 2004), p. 352.

"I will never forget that sound" (274): Jacek Fedorowicz quoted in Davies, *Rising '44,* pp. 360–61.

Chapter 32

"on the fences of all the stations" (289): Korboński, *Fighting Warsaw,* p. 406.

Chapter 34

Captions read: "dead city," "a wilderness of ruins," "mountains of rubble" (295): archival photographs reproduced in Davies, *Rising '44.*

Chapter 35

"half a million at most" (299): Joseph Tenenbaum, *In Search of a Lost People: The Old and New Poland* (New York: Beechhurst Press, 1948), pp. 297–98.

"Anyone who dared to praise pre-war independence" (301): Davies, *Rising '44,* p. 511.

"I only did my duty" (307): Rostal, "In the Cage of the Pheasant."

Chapter 36

"the little iridescent green glossy starling" (311): Heck, *Animals,* p. 61.

"what the French call a *polémique"* (312): Herman Reichenbach, *International Zoo News,* vol. 50/6, no. 327 (September 2003).

"some of the pale pink of human skin, some golden, some blue-gray, all flat" (314): Bruno Schulz, *The Street of Crocodiles,* trans. Celina Wieniewska (New York: Penguin Books, 1977), pp. 27–28.

In 2003, Magdalena Gross's sculpture *Chicken* was auctioned by the Piasecki Foundation to help raise money for autism research in Poland.

Bibliography

Aly, Götz, Peter Chroust, and Christian Pross. *Cleansing the Fatherland: Nazi Medicine and Racial Hygiene*. Baltimore, Md.: Johns Hopkins University Press, 1994.

Beebe, C. William. *The Bird: Its Form and Function*. Photos by Beebe. New York: Henry Holt, 1906.

Block, Gay, and Malka Drucker. *Rescuers: Portraits of Moral Courage in the Holocaust*. Prologue by Cynthia Ozick. Revised ed. New York: TV Books, 1998.

Calasso, Roberto. *The Marriage of Cadmus and Harmony*. Translated by Tim Parks. New York: Vintage Books, 1994.

Cooper, Rabbi David A. *God Is a Verb: Kabbalah and the Practice of Mystical Judaism*. New York: Riverhead Books, 1998.

Cornwell, John. *Hitler's Scientists: Science, War, and the Devil's Pact*. New York: Penguin Books, 2004.

Davies, Norman. *God's Playground: A History of Poland*. Vol. 1, *The Origins to 1795*. New York: Oxford University Press, 2005.

———. *Heart of Europe: The Past in Poland's Present*. New York: Oxford University Press, 2001.

———. *Rising '44: The Battle for Warsaw*. London: Pan Books, 2004.

Davis, Avram. *The Way of Flame: A Guide to the Forgotten Mystical Tradition of Jewish Meditation.* New York: HarperCollins, 1996.

Deichmann, Ute. *Biologists Under Hitler.* Translated by Thomas Dunlop. Cambridge, Mass.: Harvard University Press, 1996.

Ficowski, Jerzy, ed. *Letters and Drawings of Bruno Schulz: With Selected Prose.* Translated by Walter Arndt with Victoria Nelson. Preface by Adam Zagajewski. New York: Harper & Row, 1988.

———. *Regions of the Great Heresy: Bruno Schultz, a Biographical Portrait.* Translated and edited by Theodosia Robertson. New York: W. W. Norton, 2003.

Fogelman, Eva. *Conscience and Courage: Rescuers of Jews During the Holocaust.* New York: Anchor Books, 1994

Fox, Frank. "Zagrożone gatunki: Żydzi i żubry (Endangered Species: Jews and Buffalo)," *Zwoje,* January 29, 2002.

Glass, James M. *"Life Unworthy of Life": Racial Phobia and Mass Murder in Hitler's Germany.* New York: Basic Books, 1997.

Goodrick-Clark, Nicholas. *The Occult Roots of Nazism: Secret Aryan Cults and Their Influence on Nazi Ideology.* New York: New York University Press, 2004.

Greenfield, Amy Butler. *A Perfect Red: Empire, Espionage, and the Quest for the Color of Desire.* New York: HarperCollins, 2005.

Grynberg, Michał, ed. *Words to Outlive Us: Eyewitness Accounts from the Warsaw Ghetto.* Translated and introduction by Philip Boehm. London: Granta Books, 2003.

Gutman, Israel. *Resistance: The Warsaw Ghetto Uprising.* New York: Houghton Mifflin, 1994.

Hale, Christopher. *Himmler's Crusade: The Nazi Expedition to Find the Origins of the Aryan Race.* Hoboken, N.J.: Wiley, 2003.

Heck, Lutz. *Animals—My Adventure.* Translated by E. W. Dickies. London: Methuen, 1954.

Heston, Leonard L., and Renate Heston. *The Medical Casebook of Adolf Hitler: His Illnesses, Doctors and Drugs.* London: William Kimber, 1979.

Hoffman, Eva. *Lost in Translation: A Life in a New Language.* New York: Penguin Books, 1990.

Iranek-Osmecki, Kazimierz. *He Who Saves One Life.* Foreword by Joseph Lichten. New York: Crown, 1971.

Kater, Michael. *Doctors Under Hitler.* Translated by Thomas Dunlap. Chapel Hill: University of North Carolina Press, 1989.

Kisling, Vernon, and James Ellis. *Zoo and Aquarium History: Ancient Animal Collections to Zoological Gardens.* Boca Raton, Fl.: CRC Press, 2001.

Kitchen, Martin. *Nazi Germany at War.* New York: Longman Publishing, 1995.

Klajman, Jack, and Ed Klajman. *Out of the Ghetto.* London: Vallentine Mitchell, 2000.

Knab, Sophie Hodorowicz. *Polish Customs, Traditions, and Folklore.* New York: Hippocrene Books, 1996.

———. *Polish Herbs, Flowers & Folk Medicine.* Revised ed. New York: Hippocrene Books, 1999.

Korbonski, Stefan. *Fighting Warsaw: The Story of the Polish Underground State, 1939–1945.* Translated by F. B. Czarnomski. Introduction by Zofia Korbonski. New York: Hippocrene Books, 2004.

Korczak, Janusz. *Ghetto Diary*. Introduction by Betty Jean Lifton. New Haven: Yale University Press, 2003.

Krall, Hanna. *Shielding the Flame: An Intimate Conversation with Dr. Marek Edelman, the Last Surviving Leader of the Warsaw Ghetto Uprising*. Translated by Joanna Stasinska and Lawrence Weschler. New York: Henry Holt, 1986.

Kühl, Stefan. *The Nazi Connection: Eugenics, American Racism, and German National Socialism*. New York: Oxford University Press, 1994. (*Times Literary Supplement*, August 5, 1994)

Lemnis, Maria, and Henryk Vitry. *Old Polish Traditions: In the Kitchen and at the Table*. New York: Hippocrene Books, 2005.

Lifton, Robert J. *The Nazi Doctors: Medical Killing and the Psychology of Genocide*. New York: Basic Books, 1986. (*New York Times Book Review*, September 25, 1986)

Lorenz, Konrad. "Durch Domestikation verursachte Störungen artewigenen Verhaltens." *Zeitschrift für angewandte Psychologie und Charakterkunde*, vol. 59 (1940): pp. 2–81.

Macrakis, Kristie. *Surviving the Swastika: Scientific Research in Nazi Germany*. New York: Oxford University Press, 1993.

Matalon Lagnado, Lucette, and Sheila Cohn Dekel. *Children of the Flames: Dr. Josef Mengele and the Untold Story of the Twins of Auschwitz*. New York: William Morrow, 1991.

Mazor, Michel. *The Vanished City: Everyday Life in the Warsaw Ghetto*. Translated by David Jacobson. New York: Marsilio Publishers, 1993.

Milosz, Czeslaw, ed. *Postwar Polish Poetry*. 3rd ed. Berkeley: University of California Press, 1983.

Oliner, Samuel P., and Pearl Oliner. *The Altruistic Personality:*

Rescuers of Jews in Nazi Europe. New York: Free Press, 1988. (*New York Times Book Review,* September 4, 1988)

Paulsson, Gunnar S. *Secret City: The Hidden Jews of Warsaw, 1940–1945.* New Haven, Conn.: Yale University Press, 2002.

Polen, Nehemia. *The Holy Fire: The Teachings of Rabbi Kalonymus Kalman Shapira, the Rebbe of the Warsaw Ghetto.* Lanham, Md.: Rowman & Littlefield, 1994.

Proctor, Robert. *Racial Hygiene: Medicine Under the Nazis.* Cambridge, Mass.: Harvard University Press, 1988. (*New York Times Book Review,* August 21, 1988)

Read, Anthony. *The Devil's Disciples: Hitler's Inner Circle.* New York: W. W. Norton, 2005.

Righteous Among Nations: How Poles Helped the Jews, 1939–1945. Edited by Władysław Bartoszewski and Zofia Lewin. London: Earlscourt Publications Ltd., 1969.

Robertson, Jenny. *Don't Go to Uncle's Wedding: Voices from the Warsaw Ghetto.* London: Azure, 2000.

Rostal, Jan E. "In the Cage of the Pheasant." *Nowiny i Courier,* October 1, 1965.

Schulz, Bruno. *The Street of Crocodiles.* Translated by Celina Wieniewska. Introduction by Jerzy Ficowski. Translated by Michael Kandel. New York: Penguin Books, 1977.

Sliwowska, Wiktoria, ed. *The Last Eyewitnesses: Children of the Holocaust Speak.* Translated and annotated by Julian and Fay Bussgang. Postscript by Jerzy Ficowski. Evanston, Ill.: Northwestern University Press, 2000.

Speech to SS officers, April 24, 1943, Kharkov, Ukraine. Reprinted in United States Office of Chief of Counsel for the Prosecution of Axis Criminality, *Nazi Conspiracy and*

Aggression, vol. 4, pp. 572–578, 574. Washington, D.C.: United States Government Printing Office, 1946.

Styczński, Jan. *Zoo in Camera.* Photographs. Text by Jan Żabiński. Translated by Edward Beach Moss. London: Murrays Sales and Service Co., n.d.

Szymborska, Wislawa. *Miracle Fair: Selected Poems of Wislawa Szymborska.* Translated and notes by Joanna Trzeciak. Foreword by Czeslaw Milosz. New York: W. W. Norton, 2001.

Tec, Nechama. *When Light Pierced the Darkness: Christian Rescue of Jews in Nazi-Occupied Poland.* New York: Oxford University Press, 1986.

Tenenbaum, Joseph. *In Search of a Lost People: The Old and New Poland.* New York: Beechhurst Press, 1948.

Tomaszewski, Irene, and Tecia Werbowski. *Zegota: The Rescue of Jews in Wartime Poland.* Montreal, Canada: Price-Patterson Ltd., 1994.

Ulrich, Andreas. "Hitler's Drugged Soldiers." *Spiegel* online, May 6, 2005.

Wex, Michael. *Born to Kvetch: Yiddish Language and Culture in All of Its Moods.* New York: St. Martin's Press, 2005.

Wiedensaul, Scott. *The Ghost with Trembling Wings: Science, Wishful Thinking, and the Search for Lost Species.* New York: North Point Press, 2002.

Wiesel, Elie. *After the Darkness: Reflections on the Holocaust.* New York: Schocken Books, 2002.

Żabińska, Antonina. *Ludzie i zwięrzata* (People and Animals). Warsaw: Czytelnik, 1968.

———. "Rysie" in *Nasz dom w ZOO* (Our House in the Zoo). Warsaw: Czytelnik, 1970.

Żabiński, Jan. "Relacja . . . (A Report . . . a personal reminiscence of Dr. Jan Żabiński deposited with the Jewish Historical Institute after World War II)," no. 5704, n.d. Reprinted in *Biuletyn Żydowskiego Instytutu Historycznego w Polsze* (Warsaw), no 5. 65–66 (1968).

Zaloga, Steven J. *Poland 1939: The Birth of Blitzkrieg.* Oxford: Osprey Publishing, 2002.

Zamoyski, Adam. *The Polish Way: A Thousand Year History of the Poles and their Culture.* New York: Hippocrene Books, 2004.

Index